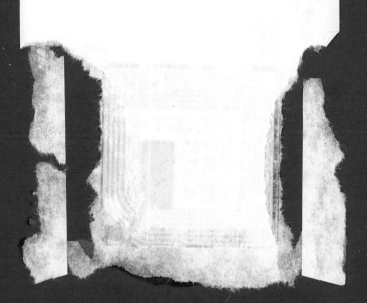

///// DRIVING FORCE

DRIVING FORCE ///////////////////
THE NATURAL MAGIC OF MAGNETS

James D. Livingston

Harvard University Press

Cambridge, Massachusetts, and London, England

Library of Congress Cataloging-in-Publication Data

Livingston, James D., 1930–
    Driving force : the natural magic of magnets /
    James D. Livingston.
        p.   cm.
    Includes bibliographical references and index.
    ISBN 0-674-21644-X (alk. paper)
    1. Magnets.   2. Magnets—Industrial applications.
    I. Title.
    QC757.L58   1996
    538′.4—dc20   95-39595
    CIP

Designed by Gwen Frankfeldt

Second printing, 1996

To the four points of my personal compass—
Sherry, Joan, Susan, and Barbara

///// CONTENTS

 # PREFACE

I recently told a friend that I was writing a book on magnets. "Oh," he said, "a children's book."

This book is not for children. I don't mean that it's X-rated, or even R-rated (maybe PG?). Actually, young children might enjoy some of the pictures, and teens and subteens with curiosity about the high-tech world around them might enjoy learning how magnets provide them with much of their music and entertainment. But I wrote this book primarily for adults.

My friend's first impression that a book on magnets was for children was not without some foundation, since many children's science books do deal with magnets. (One recent title I enjoyed was *Mudpies to Magnets: A Preschool Science Curriculum.*) Toys made with magnets are common, and many adults—including Albert Einstein and myself—have childhood memories of playing with magnets. The mysterious forces between magnets can fascinate even the very young. But for many of my *adult* years, General Electric paid me a good salary to work on magnets used in their products—motors and meters, transformers and transducers. Magnets are for children *and* for adults.

My friend then became concerned that my topic was too narrow. "OK, they're good for children's toys and for holding notes to refrigerators," he said, "but how many chapters can you get out of that?"

Magnets do a lot more than hold notes to refrigerators. A multitude of hidden magnets provide, both literally and figuratively, the "driving force" of much of today's technology. The forces that drive the motors in our appliances, that generate music in our earphones and CD players, that form images on our TV screens, that store information in our computers, and that deliver electricity to our homes to operate all these technological wonders are all provided by magnets.

"But magnets have been around for hundreds of years," he said. "Why not write about some hot new topic, like superconductors?"

But I WILL be writing about superconductors. Today's magnets are much stronger than yesterday's magnets, and the strongest are made with superconductors. Modern materials science has produced vastly improved magnetic and superconducting materials, which in turn have led to remarkable advances in technology. And the topic of magnets spans many areas of interest. This book deals with:

- astronomy (magnetic fields of planets, sun, stars),
- biology (magnetic bacteria, magnetic fields of the brain),
- geology (magnetic rocks and continental drift),
- medicine and health ("animal magnetism," magnetic resonance imaging, power lines and cancer),
- technology (computers, "maglev" trains, credit cards),
- physics (the structure of matter, the effects of magnetic and electric fields, particle accelerators, relativity),
- warfare (magnetic mines, radar, the A-bomb),

- entertainment (toys, games, magic),
- pseudoscience (magnetotherapy, ESP, dowsing),
- literature (Plato, *Gulliver's Travels, Tales of the Arabian Nights*),
- theater (Ben Jonson, Gilbert and Sullivan),
- music (Mozart, Madonna),
- movies (James Bond, *Star Trek*),
- and even comics (Dick Tracy, Magneto).

Readers will learn about the magnets that guided Columbus (and luckily misguided him a bit), the magnets that Mesmer used to mesmerize eighteenth-century Paris (but failed to fool Benjamin Franklin), and the magnets that helped AC power defeat DC power (despite the execution of many animals and one man in the DC cause). You'll learn about the magnetic device called "the most valuable cargo ever brought to our shores," which defeated Hitler's U-boats and now heats our morning coffee. You'll learn that magnets are not only *on* your refrigerator, they are *in* your refrigerator, and that magnetic refrigeration has been used by scientists to produce record-low temperatures. And you'll learn how to climb walls and ceilings like Spiderman with magnetic "Grippers." Like me, you may no longer be a child, but you'll find that magnets still provide lots of wonder and lots of fun.

Before we start, I feel I should explain why I've used the word *magic* in the title of a book about science and technology. Most dictionaries define magic in three different ways. One kind of magic is the art of entertaining with illusions and sleight of hand; although magnets play a role in that art, that's not the kind of magic I'm referring to. Another kind of magic is the unscientific realm of the supernatural; although much supernatural thinking has been applied to magnets over the centuries, that's not the kind of magic I mean either. A third definition of magic applies to qualities and objects that are mysterious and awe-inspiring even though we know they are real and not supernatural. Magnets are one of nature's wonders that possess that quality. In that sense, this book is about the many fascinating facets of the "natural magic" of magnets.

///// DRIVING FORCE

///// A MAGICAL FORCE

1

The world is full of wonder to a young child. When young Albert Einstein first encountered a magnetic compass, he was intrigued to find that, no matter which way he turned it, some hidden force turned the compass needle to point north–south. He recalled in his autobiography:

> A wonder of such nature I experienced as a child of 4 or 5 years, when my father showed me a compass. That this needle behaved in such a determined way did not at all fit into the nature of events which could find a place in the unconscious world of concepts (effects connected with direct "touch"). I can still remember—or at least believe I can remember—that this experience made a deep and lasting impression upon me. Something deeply hidden had to be behind things.

Me too. One of the few clear memories I have of my own childhood is of exploring the mysterious forces acting between magnets. I recall playing with two toy dogs, one a black Scottish terrier and the other a white Highland terrier, each about an inch long and with its four feet firmly planted on a powerful inch-long magnet. When I held one dog and moved it toward the other, head to head, it would eventually come up against an invisible barrier and I would feel it being pushed away. Suddenly, the other dog would spin around and attach itself, tail to head, to the dog I was holding. (I now realize that real dogs often get acquainted in this way.) Head to head or tail to tail, the dogs repelled each other, but head to tail, a strong attractive force pulled them together.

I could feel the attractive and repulsive forces whenever the dogs got near each other—well before they touched. The magnetic forces acted even when they were separated by a thin barrier, like a comic book. With one dog on top of the book, I could move the other dog under the book and the "top dog" would follow, seemingly on its own. It was magical indeed. Little did I know then (about 1940) that I would spend much of my adult life at General Electric helping to develop magnets far stronger than those on which my two terriers were mounted.

What seems most magical about magnets is that they exert forces *without touching.* This wonder of "force at a distance" has puzzled philosophers and scientists from ancient times to the present day. Engineers have learned how to harness the magnetic force to create many of the wonders of modern technology, and physicists have developed models that allow very precise force calculations that are in excellent agreement with experiment. Theorists even argue, as I discuss in Chapter 16, that the forces between magnets and the force you feel when your hand touches a chair are fundamentally the same force. Maybe so. But to most of us, the forces between magnets still seem a lot more mysterious and magical than forces between objects that touch. What delivers the force from one magnet to the other?

The motion of Einstein's compass needle is an even greater mystery. With my two terriers, it was clear that the forces were produced by the magnets, but what turned the compass needle? The earth's magnetic field did, but what is a "magnetic field," and where does the earth's magnetic field come from? And why does the earth's magnetic field turn an iron needle but not a brass needle or a wooden needle? Magnets present even more mysteries. Read on.

///// Paper Clips and Refrigerators

"Something deeply hidden" was behind the workings of Einstein's compass and my toy dogs. I soon learned that magnets also had the remarkable property of being able to pick up paper clips, nails, and other objects of iron or steel. And when touched with a magnet, paper clips themselves became magnets, able to attract other paper clips! If the magnet is strong enough, you can hang a long chain of paper clips under it, but when the magnet is removed the attractive force disappears and the clips fall apart. Einstein's compass needle and the mounts for my terriers were what we call *permanent magnets.* The paper clips in a chain become *temporary magnets,* exerting magnetic forces on other paper clips only when connected to a permanent magnet.

In most homes, the most visible permanent magnets are those in the kitchen, holding various important bits of paper—shopping lists, reminders, children's art—to the refrigerator. My refrigerator currently has eight magnets attached to it: one green turtle, one red tulip, and a variety of simple circular and rectangular shapes, some carrying advertising from local businesses. But I'm a piker compared to Marlou Freeman of Laurel, Maryland, featured recently in *People* magazine. Her collection of 2,300 magnets was exhibited in a New York art gallery, accompanied by a color catalogue of "Marlou's Magnets."

Refrigerator magnets, like the mounts for my terriers, are permanent magnets. The refrigerator itself, thankfully, is not. (Otherwise, knives, forks, and other steel objects might suddenly be drawn toward it and cause messy accidents.) It can, however, be a temporary magnet, if it is magnetized enough underneath my green turtle and red tulip to provide an attractive force. My refrigerator magnets are attached to the steel sides of my refrigerator. The front, because it is made of a plastic, is unable to provide magnetic forces. Most materials cannot become either permanent or temporary magnets—only iron, steel, and a few other things have "the right stuff." (You may protest that some refrigerator magnets look like plastic, and you're right. But their magnetism comes from powdered magnets that are imbedded in the plastic, not from the plastic itself.)

Throughout this book, we'll encounter two kinds of magnets— permanent magnets, like compass needles and refrigerator magnets, and temporary magnets, like paper clips and refrigerators. Occasionally, a paper clip or steel nail may retain a bit of attractive power even after the permanent magnet is removed and become, at least for a while, a weak permanent magnet. But the distinction between permanent and temporary magnets is much clearer than the dividing line between, say, tall people and short people, thanks to the scientists and engineers that develop magnetic materials. They've invested much time and energy in developing materials that are good permanent magnets, and in developing materials that are good temporary magnets, because both types have important applications. There are very few applications for materials that are on the borderline, and therefore most magnetic materials clearly fall in one category or the other. And just as we can quantify tallness or shortness with a single number, a person's height, there are quantities that scientists and engineers use to define the important properties of permanent and temporary magnetic materials, which will be introduced in later chapters.

///// James Bond and Jaws

Permanent magnets (Figure 1.1A), temporary magnets, and the existence of attractive and repulsive magnetic forces have been known to mankind for thousands of years. It was not discovered until the nineteenth century, however, that magnetic forces could

Figure 1.1. Magnetic field patterns *(dashed lines)*. (A) A permanent magnet. (B) An air-core electromagnet (i.e., a current-carrying coil of wire); unless the wire is superconducting, the field produced is limited by resistive heating. (C) An iron-core electromagnet; the temporary magnet inside the current-carrying coil amplifies the field produced by the coil.

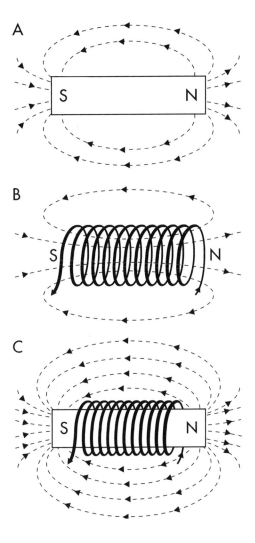

be produced simply by passing an electric current through a copper wire. If the wire is wound into a coil of many turns (like a spring or a Slinky), the coil acts like a magnet as long as current is flowing (Figure 1.1B). The stronger the current, the stronger the magnet.

Unfortunately, the amount of current you can pass through a

copper wire is limited—higher and higher currents make the wire hotter and hotter. To pass enough current to make a magnet as strong as those on my terriers would melt the copper. The solution to this problem is to put some iron (a good temporary magnet) inside the coil of wire (Figure 1.1C).

When current flows through the coil, the magnetism produced by the coil is amplified by factors of hundreds or even thousands by the iron core. Magnets of this type not only produce strong magnetic forces, but they may also be turned off and on by turning the electric current off and on. We call such magnets *electromagnets*. (An empty coil is called an air-core electromagnet.) You may remember an experiment in school where you wound a wire around an iron nail, attached the wire to a battery, and found that the nail became a strong magnet. It's a standard exercise in early science classes and is described in many science books for children. But don't be upset if you've forgotten it. You've probably seen a few electromagnets in the movies.

A powerful electromagnet played an important role in the James Bond movie *The Spy Who Loved Me*. The villain in this movie was "Jaws," a huge man who used his steel teeth to bite several unfortunate characters to death. In a scene near the end, Jaws was closing in on James Bond with evil intent when Bond noticed a large electromagnet conveniently located overhead. Bond quickly turned on the current to the electromagnet, and Jaws found himself suspended in the air, with his steel teeth firmly attached to the magnet. Bond then maneuvered the magnet, with Jaws attached, over a pool containing a hungry shark and turned off the current. The magnetic force disappeared, dropping Jaws into the pool and an encounter with the shark. (This was not the end of Jaws, however, as his bite turned out to be more deadly than the shark's. He swam safely away, to reappear later in another Bond film, *Moonraker*.)

Electromagnets are similarly used in scrapyards to lift and transport discarded automobiles to and from the machines that crush them into compact forms of scrap metal (sub-*sub*-compact cars). The ability to switch magnetic forces off and on by switching electric currents off and on is so convenient that electromagnets have found many applications beyond the lifting and transporting of villains and cars. Until recently, they've actually had more applications in twentieth-century technology than permanent magnets, but the balance is shifting as a result of the recent development of improved permanent magnets.

Magnetic forces are only one of several forces that things can exert on other things, and scientists have introduced various concepts to explain these forces, including the presence of an invisible "field" of force. One familiar force is the force of gravity, the force that makes apples fall and keeps the solar system together. Ever since Isaac Newton developed his theory of gravitation (in the late seventeenth century), we've said that masses attract each other via invisible gravitational fields. Masses are sources of gravitational fields, and they are acted on by forces from gravitational fields. What makes this theory more than just words is that Newton's mathematical model allows us to *calculate* gravitational fields and forces and predict the motion of things, in agreement with observations and measurements.

Once the existence of positive and negative electric charges began to be understood, we said that static electric charges attract and repel each other via electric fields. And once electromagnets were discovered, we said that *moving* electric charges (electric currents) attract and repel each other via *magnetic fields*. Electric currents are sources of magnetic fields and are acted on by forces from magnetic fields. And our mathematical theories allow us to calculate magnetic fields and forces that agree with experiment. Neat.

But what about permanent magnets? And what about temporary magnets like paper clips, whose magnetism remains hidden until stimulated by permanent magnets or current-carrying coils? My terrier mounts and my paper clips were also sources of magnetic fields, and were acted on by forces from magnetic fields, but there were no electric currents in them. Or were there? We'll postpone discussion of these mysteries until Chapter 6.

///// Hidden Magnets

Most magnets, even those in your home, are not as visible as refrigerator magnets. Nevertheless, they are reliably working their magic, unseen, in many household appliances. Let's suppose that you've brought *The Spy Who Loved Me* home from the video store to play on your VCR. You may be surprised to realize how many hidden magnets are working to bring James Bond and Jaws to your television screen:

The Motor: The motor driving the tape consists of two magnets, one roughly doughnut-shaped, the other rotating in the hole of the doughnut in response to magnetic forces. In some motors, one magnet is a permanent magnet and the other is an electromagnet. In others, both are electromagnets. Either way, the forces between the two magnets convert electrical energy into mechanical energy and move the tape. Motors run many other appliances—washers, dryers, refrigerators, air conditioners—all employing the attractive and repulsive forces between two magnets to do their work.

The Speaker: Most speakers consist of a movable coil within a stationary permanent magnet. Varying currents through the coil produce varying magnetic forces on the coil, resulting in vibrations of the coil and an attached diaphragm, which in turn vibrates the air. These acoustic vibrations are what you "hear" when Roger Moore says, "My name is Bond—James Bond." Radios, telephones, doorbells, etc., all generate sound from electrical inputs via magnetic forces. (If you want more details of how motors, speakers, and the next items utilize magnetic forces, good. You'll have motivation to read later chapters. Right now I'm just trying to impress you with how many hidden magnets you often use without realizing it.)

The Picture Tube: The picture on your TV screen results from variations in electric current as the current is swept across the screen. The forces that direct the horizontal and vertical sweeps of the current across the screen are produced by two pairs of electromagnets. Other magnets within your TV set play other important roles—most TV sets contain about two pounds of magnets.

The Videotape: All the sounds and pictures making up *The Spy Who Loved Me* are coded magnetically into billions and billions of tiny permanent magnets imbedded in the plastic tape. Each magnet is magnetized in one direction, or its opposite, along the tape, and in the complex pattern of these magnetic directions is coded all the sounds and pictures, including the picture of Jaws held suspended over a shark pool by an electromagnet. The pattern of magnetic directions of many tiny permanent magnets also stores information on audio cassette tapes, the floppy disks and hard disks of computers, and credit cards and ATM cards.

The Head: The pickup head is a small electromagnet that translates the magnetically coded information in the tape into electric currents and sends these currents to the speaker and picture tube. We've learned that sending currents into an electromagnet produces magnetic fields, and we send currents into the head to *record* a tape. But when we *play* a tape, the opposite occurs—magnetic fields from the moving tape produce electric currents in the head. Electricity in motion (current) produces a magnetic field, and magnetism in motion produces an electric field.

This is an important effect of magnetic fields discovered by Michael Faraday, an English scientist, in the nineteenth century. Faraday found that if a wire was exposed to *changing* magnetic fields, an electric current was "induced" in the wire. This effect is called *induction* or, more precisely, electromagnetic induction. The phenomenon of induction, allowing us to generate electric currents from changing magnetic fields, is used not only in pickup heads but also in the generators and transformers that produce and deliver electric power to the wall outlet from which we run our VCR and TV.

So the "simple" process of playing a videotape of James Bond and Jaws employs unseen magnets in a motor, a speaker, a picture tube, a pickup head, and the videotape itself, not to mention those in the electrical system delivering power to your home. Magnets do a lot more for us today than hold notes to our refrigerators!

///// Facts about the Force

In this first chapter, we've already touched on most of the basic facts about magnetic forces. Some you probably learned long ago and still remember, and some (like induction) are more complex and may be new to you. It will help us in later chapters if we review them briefly here.

Fact 1: If free to rotate, permanent magnets point approximately north-south.

Compass needles rotate easily. With other permanent magnets, you may have to hang them from a long string to allow them to swing in response to the weak north-south magnetic field of the earth. The part of the magnet that points north we call the *north pole* (or north-seeking pole), and the part that points south we call the *south pole.* (At least that's standard scientific and engineering terminology today. Some people, for reasons noted in Chapter 3, reverse this notation.)

Fact 2: Like poles repel, unlike poles attract.

As I recall, the permanent magnets on my toy terriers had north poles under their heads and south poles under their tails. Head to head or tail to tail, they repelled. Head to tail, they attracted.

Fact 3: Permanent magnets attract some things (like iron and steel) but not others (like wood or glass).

Apparently paper clips and refrigerators have the potential to *become* magnets, but their magnetism remains hidden inside them until stimulated by a permanent magnet (or a coil carrying current—see Fact 7).

Fact 4: Magnetic forces act at a distance, and they can act through nonmagnetic barriers (if not too thick).

My terriers attracted each other, and could attract paper clips, even when on opposite sides of a thin comic book. It seemed pretty magical, and still does.

Fact 5: Things attracted to a permanent magnet become temporary magnets themselves.

You can hang many paper clips, end to end, below a strong permanent magnet—but they fall apart again as soon as you remove it.

Fact 6: A coil of wire with an electric current flowing through it becomes a magnet.

The ends of the coil become north and south poles (Figure 1.1B), which reverse if you reverse the direction of the current. The stronger the current, the stronger the magnet. The more turns per inch in the coil, the stronger the magnet.

Fact 7: Putting iron inside a current-carrying coil greatly increases the strength of the electromagnet.

Iron-core electromagnets (Figure 1.1C) can be strong enough to lift villains and automobiles, and they can be turned off by turning off the current. Most electromagnets have cores of iron or similar magnetic materials.

Fact 8: Changing magnetic fields induce electric currents in copper and other conductors.

You'll recall this is how the pickup head of your VCR translates the magnetic code in the moving videotape into electric currents. This important phenomenon called *induction* also generates electric power from Niagara Falls (see Chapter 8), levitates the Japanese superconducting train (Chapter 10), and, coupled with fact 6, produces the magnetic field of the earth (Chapter 3). For many readers, this may be the least familiar of the 8 facts, but it will come up so often in later chapters that it may eventually become the most familiar. Electromagnetic induction is not only of great technological importance. As we'll see in Chapter 16, it was also influential in the development of Einstein's theory of relativity.

The first five of these eight facts about magnetic forces have been known for many hundreds of years. Facts 6 through 8, relating magnetism to electricity, have been known only since the nineteenth century. In later chapters I'll introduce two more "facts about the force," but for the next chapter, on the early history of mankind's knowledge and use of magnets, we'll need only the first five.

THE FIRST FIVE FACTS ABOUT THE FORCE

1. North poles point north, south poles point south.
2. Like poles repel, unlike poles attract.
3. Magnetic forces attract only magnetic materials.
4. Magnetic forces act at a distance.
5. While magnetized, temporary magnets act like permanent magnets.

///// ROMANCING THE STONES

2

Magnetic rocks, today called lodestones, are found in nature throughout the world. The magnetic forces between lodestones must have been noticed, and wondered at, by prehistoric men and women many thousands of years ago. Once the smelting of iron was developed (about 1200 B.C.) and the use of metallic iron became widespread, the ability of lodestones to attract iron probably also became widely known. In one of Plato's early dialogues, written about 400 B.C., he has Socrates say:

> There is a divinity moving you, like that contained in the stone which Euripides calls a magnet, but which is commonly known as the stone of Heraclea. This stone not only attracts iron rings, but also imparts to them a similar power of attracting other rings; and sometimes you may see a number of pieces of iron and rings suspended from one another so as to form quite a long chain: and all of them derive their power of suspension from the original stone. In like manner the Muse first of all inspires men herself; and from these inspired persons a chain of other persons is suspended, who take the inspiration.

Clearly the aim of Socrates here is not to explain the magical forces of magnets. He is discussing the inspiration of poets, using magnets as analogy. Apparently, he considered the permanent-magnet properties of lodestones and the temporary-magnet properties of iron (Socrates used iron rings, I used paper clips) to be common knowledge to readers of his day. This and other writings of the early Greeks show that basic facts about magnetic forces—facts 2 through 5 in the previous chapter—were widely known at least several centuries before Christ.

But what about fact 1, the north-south orienting of magnets? Surprisingly, Western literature shows no mention of this property of magnets until much later—until nearly 1200 A.D. The term *lodestone*, referring to the use of magnetic compasses for navigation (*lode* being old English for "leading" or "guiding"), didn't appear until about 1500.

Several old Greek and Roman writings claim that the word *magnet* derives from Magnes, the name of a shepherd who noticed one day that the iron nails in his shoes stuck to certain rocks. Probably not true, although a student of mine from Greece tells me this legend is still taught there as fact today. The name more likely derives from one of several locations in ancient Greece called Magnesia, where lodestones were found. Presumably lodestones were also found at Heraclea, mentioned above by Socrates (via Plato).

Wherever they were found, and whatever they were called, lodestones were considered very mysterious by scholars of ancient Greece and Rome. In *Natural History*, a comprehensive description of nature that was written in 77 A.D. and remained an authority on scientific matters until the Middle Ages, Pliny writes: "For what is more strange than this stone? In what field has Nature displayed a more perverse willfulness? She has given to rocks a voice . . . she has endowed the magnet with senses and hands."

///// Loving Stones

In China, the lodestone went by the name *tzhu shih,* the "loving stone," because of its power to attract iron. (In discussing "loving stones," some authors discreetly refer to parental love for children, but others note that *tzuh* can mean copulation, a somewhat earthier interpretation of loving.) The French, often considered experts on romance, use the word *aimant* both for "magnet" and for "loving, affectionate."

In many cultures and times, the attractive forces between magnets have been a common analogy for the attractions between loved ones. Consider Ibn Hazm, an eleventh-century Muslim scholar whose *Ring of the Dove* is considered to be the most comprehensive book in Arabic on the subject of love. He wrote:

> My eyes find nowhere else to look but at you,
> Like what happens with the lodestone (and the pieces of iron),
> Changing direction to right and left, in accordance with where you are.

Another example is *The Magnetic Lady,* written by Ben Jonson, a dramatist whose popularity in seventeenth-century England was second only to Shakespeare's. The lady of the title was Lady Loadstone, who in the play successfully attracted Captain Ironside and strongly influenced characters named Compass and Needle. Today we still speak of magnetic personalities and, by extension, apply the term *magnet* to anything that attracts. The Grand Canyon is said to be a "magnet for tourists," and we even have "magnet schools" designed to attract students (and perhaps also government funding).

But back to China. Joseph Needham, the foremost Western scholar of Chinese science and civilization, writes that the study of magnetism "was the greatest Chinese contribution to physics." Whereas the north-south orientation of magnets was apparently unrecognized in Europe until nearly 1200, the Chinese had developed a compass-like device, held as a secret of magicians of the emperor's court, over a thousand years earlier!

That device was a spoon carved from a loving stone and placed on a polished bronze plate, where it was free to rotate in response to the earth's field. In 1948, a Chinese scholar named Wang

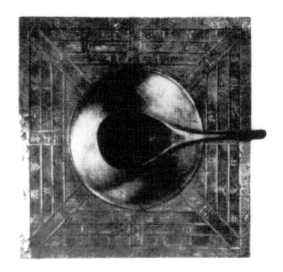

Figure 2.1 Reproduction of an ancient Chinese spoon compass. The rounded bottom of the spoon-shaped lodestone would rotate easily on the brass "earth plate" until the spoon handle pointed south. Compasses like this were used in China over two thousand years ago.

Chen-To presented convincing evidence that such a spoon had been used as a south-pointing device in China as early as 200 B.C. A modern model of this device (Figure 2.1) was presented to me and to other participants in an international conference on magnets held in Beijing in July 1992. The spoon shape of the stone was believed to represent the constellation Ursa Major, the "Big Dipper."

There is no evidence that the Chinese used the lodestone spoon for navigation. It was probably put to more magical ends in the ancient occult practice known as geomancy, in which good fortune could be assured by accurate alignment of houses, beds, and other objects with the heavens. By at least the sixth century A.D., the Chinese had learned how to transfer the directive property of the lodestone to small pieces of iron, which could be floated on water if suitably shaped. Today we would say the lodestone was used to "magnetize" the iron. The first use of magnetized iron needles for navigation probably occurred in China two or three centuries before they became known in Europe. Once the magnetic compass came into common use, however, Europeans used it to great effect to explore, dominate, and plunder large portions of the globe.

Lodestones were used to magnetize iron compass needles for many centuries, until they were superseded by steel permanent magnets in the eighteenth century and electromagnets in the nineteenth century. Old lodestones used to magnetize compass needles are now just museum curiosities. I recently purchased some fist-sized lodestones from a science supply house. To demonstrate the magnetic nature of the stones, they were apparently dipped in iron filings before shipping. They look like rocks with a three-day beard.

///// Inside the Loving Stones

What are lodestones made of? How strong are they? It is only in the twentieth century that we have accurately measured the magnetic properties of lodestones. And now we have optical and electron microscopes to study their internal structure and determine the source of their permanent-magnet properties.

Most geology books say that lodestones are made of *magnetite*, a compound of iron (Fe) and oxygen (O) with the formula Fe_3O_4. Most of the earth's surface consists of oxides—compounds of various elements with oxygen—and iron is the third most common of these elements (behind silicon and aluminum). Iron forms other oxides: FeO (wustite) and two varieties of Fe_2O_3 (hematite and maghemite). Although hematite is the most common of the iron oxides, magnetite is the most magnetic. It seems reasonable to conclude that lodestones are made of magnetite—and they are, mostly.

But there's a bit more to it than that. If you make a nice pure, homogeneous sample of magnetite in your laboratory, you'll find that it does *not* act like a lodestone. It can act as a *temporary* magnet (though weaker than steel paper clips and refrigerators), but it does not act as a *permanent* magnet. Apparently lodestones are not simply pure magnetite.

If you polish a lodestone until you obtain a smooth surface, and look at it under a microscope at magnifications of $1,000\times$ or higher, you can see that it is not homogeneous, that it is not all magnetite. Lodestones that contain only iron and oxygen are mostly magnetite but also contain some tiny regions of maghemite. Other lodestones contain a few percent of other elements, like titanium, aluminum, or magnesium, and the microscope shows that these elements are not uniformly distributed inside the stone. There are tiny regions that contain more of these other elements than the rest of the stone. Physicists now realize that these tiny regions are important. This nonhomogeneous distribution of elements that we can see only under a microscope—this "microstructure"—is what makes a lodestone a permanent magnet. The tiny regions of maghemite within the magnetite, or the tiny regions richer in other elements, are necessary to make magnetite into a lodestone—into a loving stone.

Similarly, pure iron that is homogeneous under the microscope, although a great temporary magnet, is not a permanent magnet. But if the iron isn't pure, and some of the additional elements collect into microscopic regions within the iron, the iron can become a weak permanent magnet—not as good as a lodestone, but "permanent" enough to serve as a compass needle. It was known centuries ago that the best compass needles were made of steel—iron hardened by the presence of carbon and, sometimes, other elements.

To clarify the difference between lodestones and iron, it's helpful to define two quantities used to describe the strength of a magnet. The first measures how much magnetic field the material is capable of producing if it is magnetized by a strong field from another source—say a current-carrying coil. This quantity, the *saturation magnetization* of the material, represents the ultimate magnetic field that the magnet, temporary or permanent, can produce. For both temporary and permanent magnets, a high saturation magnetization is desirable. It will determine, for example, how much steel a magnet of fixed size will be able to lift.

The second important quantity measures how "permanent" the magnet is—how much field in the opposite direction is needed to remove its magnetization, to "demagnetize" it. This is called the *coercivity* of the material. An ideal temporary magnet has a coercivity of nearly zero, so that it loses most or all of its magnetization once the magnetizing field is removed. An ideal permanent magnet instead has a high coercivity—is difficult to demagnetize.

Fortunately, we can use the same unit to describe both the saturation magnetization and the coercivity (and the strength of magnetic fields). The unit we will use is the *gauss*, named after German mathematician and physicist Carl Friedrich Gauss (1777–1855). A magnetic field of one gauss is about twice the average magnetic field at the earth's surface.

I should warn you that several other units for magnetic quantities are used in the scientific literature. Of growing use is the tesla, named after Nikola Tesla (1856–1943), Serbian-American inventor and researcher in electromagnetism. One tesla is equal to 10,000 gauss or 10 kilogauss. Magnetics experts also prefer to use different units for magnetization and magnetic field, but that complication seems unnecessary here.

The saturation magnetization of iron is over 20 kilogauss, but that of a lodestone is typically less than 4 kilogauss. Thus the ultimate magnetic field that can be produced by magnetized iron is much more than what can be produced by a lodestone. Iron is a great *temporary* magnet.

Lodestones are the clear winners in the coercivity category, however. Their coercivities are typically about 200 gauss, while those of pure iron are usually one gauss or less. This is why a lodestone is a better *permanent* magnet than pure iron. Still, as noted above, a little carbon in the iron can raise the coercivity to a few gauss and make it "permanent" enough in the earth's field of one-half gauss to serve as a good compass needle.

The connection between coercivity and microstructure—those tiny regions of different chemical composition within permanent magnets that make them hard to demagnetize—was a focus of my research at General Electric for several years. In fact, the connection between the microstructure of materials and macroscopic properties is a major component of today's science of materials. Magnetic coercivity is only one of many properties of materials that are very sensitive to microstructure. But I'll save explaining the microstructure-coercivity connection for Chapter 6.

Lodestones have low saturation magnetizations but high coercivities. They are permanent magnets but can't produce large magnetic fields. Iron, with its high saturation magnetization but low coercivity, is a great temporary magnet but a poor permanent magnet. By the sixteenth century, the two materials had been combined to produce a permanent magnet stronger than either material separately could be. Lodestones with iron attachments, called "armed" lodestones (Figure 2.2), became the strongest permanent magnets available.

Armed lodestones provide an example of a principle often used by today's materials scientists and engineers. All materials have some strengths and some weaknesses. Often an intelligent

Figure 2.2 Armed lodestone from the time of Sir Francis Drake (late sixteenth century). It is about 2.5 inches on a side and has a brass casing and two iron pole pieces at top and bottom. During long sea voyages, the iron pole pieces, magnetized by the field from the lodestone, could be used to remagnetize compass needles by touch whenever necessary. This piece is now on display at the National Maritime Museum in Greenwich, England.

combination of two materials can utilize the strengths of each to provide a "composite" material superior to its separate components. As with romance, two can be better than one.

One interesting question about lodestones remains unresolved. They were used, first by the Chinese, later by the Europeans, to magnetize iron compass needles. But what magnetized the lodestones? We do know that when rocks form, from cooling lava or from sedimentation, they may become magnetized simply by the earth's magnetic field. As we'll see in the next chapter, this happens in nearly all magnetic rocks, and the magnetic orientation of geological formations has been used to learn a lot about the history of the earth. However, many lodestones are magnetized more strongly than would be expected simply from the earth's field, which is only about half a gauss. The best explanation so far is that they were magnetized by lightning strikes—by the large magnetic fields associated with the huge currents carried in lightning. It's a romantic concept: humble rocks of magnetite transformed into lodestones—into loving stones—by a bolt from the blue.

///// Romancing from Afar

The mysterious response of a magnetized iron needle to the "invisible hand" of the earth's magnetic field has inspired many fancies. What has become known as the "sympathetic telegraph" was first described in Giambattista della Porta's *Natural Magic,* published in Naples in 1558. This imaginary device consisted of two compasses, with their needles magnetized by the same lodestone. The compass dials, rather than showing directions, were labeled with the letters of the alphabet. The movement of one needle supposedly caused the identical movement of the other, allowing a message to be sent, letter by letter, from afar. And this three hundred years before Samuel Morse and Western Union!

This concept of a magnetic telegraph appeared often in later literature, sometimes presented as fact, sometimes debunked as fable (Galileo was among the debunkers). In 1711, a distraught wife wrote *The Spectator,* a London periodical, for advice on how to bear the long absence of her beloved and devoted husband. The essayist Joseph Addison, acting as an early Ann Landers, first responded with some practical advice. He then wished that some magician could present a pair of Porta's synchronized magnetic needles to separated lovers, enabling them to communicate with each other, even "when they were guarded by spies and watches, or separated by castles and adventures." He also made the practical suggestion that, in addition to the alphabet, the "lover's dial-plate" should contain "several entire words which always have a place in passionate epistles." The message "I love you," for example, would then require only three needle settings instead of eight.

The "natural magic" of magnets was never powerful enough

to allow construction of the device envisioned by Porta. The discovery of electromagnets in the nineteenth century, however, led to long-distance communication via magnetic forces in the electric telegraph and telephone. And today we can contact lovers (and others) with cellular phones, faxes, or e-mail via the Internet. Our modern magnetic means of communication are real, but they seem as magical to me as Porta's imaginary invention seemed to Famianus Strada, a seventeenth-century poet:

> Thus, if at Rome thy hand the steel applies,
> Tho' seas may roll between, or mountains rise,
> To this some sister needle will incline,
> Such nature's mystic pow'r, and dark design!

///// Looking for Lodestones

Lodestones are found in magnetite-rich ore deposits in many parts of the world. Large concentrations of magnetite, maghemite, or other strongly magnetic rocks result in local variations in magnetic field called "magnetic anomalies." For centuries, compass-like devices have been used to search for such ores, and by the eighteenth century these evolved into the "Swedish mining compass," a device that contained a magnetic needle in a low-friction jewel mounting that allowed it to rotate freely both horizontally and vertically. This and other magnetic-anomaly detectors are now known by their acronym, MADs. MADs helped the Russians locate a magnetite-rich mountain—now named Mt. Magnitnaya—near the Ural River. To exploit these iron-ore deposits, they founded a city along the river and constructed one of the world's largest iron- and steelworks. The city was named Magnitogorsk, a "Magnet City" near the "Magnet Mountain."

The introduction of airplanes greatly increased the speed with which magnetic prospecting could be accomplished. The first use of MADs for prospecting from the air was in 1936 by A.

Logachev, a Russian geophysicist studing magnetic anomalies near the city of Kursk. His detector was only sensitive enough, however, to recognize variations in magnetic field larger than one-hundredth of a gauss, about 2 percent of the earth's average field. A magnetic-anomaly detector a thousand times more sensitive was developed during World War II to locate submarines and has since been used widely to prospect for minerals and oil.

Airborne MADs, often suspended below and behind an airplane at the end of a 100-foot cable, are capable of scanning several thousands of square miles per month. With this technique, large magnetic iron-ore deposits have been found in many states and several Canadian provinces. One steel company reported in the 1960s that just a few years of data from airborne MADs had enabled the company to locate a 75-year reserve of iron ore. Magnetic-anomaly data are also of great interest to geologists, and the results from MAD surveys using airplanes, ships, and NASA's MAGSAT satellite (launched in October 1979) have been used to produce magnetic-anomaly maps of much of the earth's surface.

MADs have also been towed behind ships across much of the Atlantic and Pacific oceans. Magnetic-anomaly maps revealing the magnetization of rocks on the ocean floor have provided geologists with key evidence to develop current theories of continental drift and plate tectonics, as we will discover in the next chapter. In recent years, loving stones and other magnetic rocks have told us a lot more about our earth than which way is north.

///// A Magnetic Love Song

It seems appropriate to close this chapter on loving stones with a metaphor—that is, a song about unrequited magnetic attraction. It's from *Patience,* one of Gilbert and Sullivan's most successful comic operas. (*Patience* has a second connection with magnets.

Figure 2.3 "The magnet and churn" mentioned in Gilbert and Sullivan's *Patience* (1881), an illustration attributed to William Gilbert himself. The familiar horseshoe shape of the magnet is well suited to the low-coercivity steel magnets of that day, but this shape is not used for the high-coercivity magnets of today.

It debuted in 1881 in London's Savoy Theatre, the first in the world to employ electric lights—the electricity being generated, of course, by magnets and magnetic induction, making good use of fact 8.)

The plot in *Patience* revolves around the difficult but all-too-common circumstance of loving someone who loves another. A group of maidens all love Grosvenor, a poet who cannot return their love because he loves Patience, who in turn cannot love him because she is loyal to Bunthorne. Early in Act II, Grosvenor says to the maidens, "I know that I am loved by you, but I never can love you in return, for my heart is fixed elsewhere! Remember the fable of the Magnet and the Churn!" (Figure 2.3 shows William Gilbert's illustration of the tale.) The maidens claim ignorance of the fable, so he sings it to them:

> A magnet hung in a hardware shop
> And all around was a loving crop
> Of scissors and needles, nails and knives,
> Offering love for all their lives;

But for iron the magnet felt no whim,
Tho' he charmed iron, it charmed not him,
From needles and nails and knives he'd turn,
For he'd set his love on a Silver Churn!

His most aesthetic, very magnetic
Fancy took this turn—
"If I can wheedle a knife or a needle,
Why not a Silver Churn?"

And Iron and Steel expressed surprise,
The needles open'd their well-drill'd eyes,
The pen-knives felt "shut up," no doubt,
The scissors declar'd themselves "cut out,"
The kettles they boiled with rage, 'tis said,
While ev'ry nail went off its head,
And hither and thither began to roam,
Till a hammer came up and drove them home.

While this magnetic, peripatetic
Lover he lived to learn,
By no endeavor can magnet ever
Attract a Silver Churn!

For Grosvenor, at least, the play has a happy ending—he wins the love of Patience, his own "silver churn." But lyricist William Gilbert had his physics right: a magnet cannot attract silver (see fact 4). That fact and others were also well known to another William Gilbert, whose work nearly 300 years earlier set the foundation for the modern science of magnetism. We meet him in the next chapter.

///// MAGNUS MAGNES

3

///// The Great Magnet

Einstein's childhood observations with a compass taught him that "something deeply hidden had to be behind things." Magnets exert forces on each other, we say, via invisible magnetic fields. But what magnet was producing the magnetic field that exerted forces on Einstein's compass needle? That question was answered in 1600 by William Gilbert, physician to Queen Elizabeth I of England, in his remarkable book *De Magnete*. By careful experimentation and rigorous analysis, Gilbert concluded *"magnus magnes ipse est globus terrestris"* (the earth's globe itself is a great magnet).

De Magnete is considered by many to be the world's first great scientific treatise and a classic example of the application of "the scientific method." Gilbert argues in his preface that "stronger reasons are obtained from sure experiments and demonstrated arguments than from probable conjectures and the opinions of philosophical speculators." Galileo wrote of him, "I extremely praise, admire and envy this author, for that a conception so stupendous could come into his mind. I think him moreover worthy of extraordinary applause for the many new and true

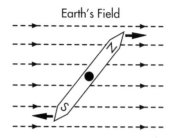

Earth's Field

Figure 3.1 Magnetic forces acting on a compass needle. Because the needle can swing freely only in the horizontal plane, it responds only to the horizontal component of the earth's field, which is uniform over the length of the needle. The forces exerted on the north and south poles of the needle are therefore equal but opposite in direction; as a result, they produce a torque that rotates the needle until it becomes parallel to the horizontal component of the field.

observations he made." Gilbert's book went into three editions, and the library of the University of Cambridge, England, has a copy of the third edition that belonged to playwright Ben Jonson. It was probably used by Jonson in writing *The Magnetic Lady,* since his pencilled notes abound in the margins of the early chapters.

In his book, Gilbert heatedly debunked various legends and superstitions that had arisen about magnets, including claims that a lodestone rubbed with garlic would not attract iron, and that a lodestone placed under the head of a sleeping woman would drive her out of bed if she were an adulteress. (The former legend he specifically disproved by experiment, but he did not make clear what experiments he did to disprove the latter.) His most important experiments, however, were observations of the orientation of iron needles placed at various positions around a lodestone carved into a sphere, observations that convinced Gilbert that the earth itself is a magnet.

First, let's recall what the compass tells us about the earth's magnetic field. The compass is constructed so that the needle, fixed at its center, can swing freely in a horizontal plane. The earth's field exerts equal and opposite forces on the north and south poles of the needle (Figure 3.1), causing it to rotate until it points roughly north-south, in which position the torque (rotational force) on the needle disappears. By responding to the torque, the compass needle shows us, *within the horizontal plane,* the direction of the earth's magnetic field.

Shortly before 1600, an English instrument maker named Robert Norman built a compass-like device that revealed new information. Called a dip needle, this device was constructed so that an iron needle could swing freely in a *vertical* plane. The needle was found to dip away from the horizontal, showing that the earth's magnetic field also had a vertical component.

Now to Gilbert's experiment with the spherical lodestone (Figure 3.2). He found that an iron needle pointed perpendicular to the surface of the sphere at the magnetic poles of the lodestone and parallel to the surface at its magnetic equator; in between, the needle was inclined to the surface. Seeing the similarity of this behavior with the behavior of dip needles in the earth's magnetic field, he made a huge (and correct) leap in reasoning and concluded that the earth was a spherical magnet with magnetic poles near the geographic poles. One practical application of this important conclusion was the use of the angle of dip, today called "inclination," as a measure of earth latitude. The greater the dip angle, the farther you are from the equator. If your dip needle points straight down, you've reached one of the earth's magnetic poles. And it's cold outside, since you're either

MAGNETIC PUSH AND PULL

The magnetic forces in Figure 3.1 cause the needle to rotate, but they do not attract the compass as a whole, southward or northward, because the forces on opposite poles are equal and opposite, yielding no *net* force. On the other hand, if the magnetic field varied with position, so that the field acting on one pole was different from that acting on the other, the two opposing forces would not be exactly equal and a net force would result. A magnetic field exerts a *net* pull or push on a magnetic material only if the field varies with distance. Iron is attracted to a magnet because the field is higher near the magnet than at a distance from it.

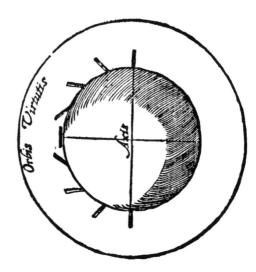

Figure 3.2 Orientation of an iron needle at various locations around a spherical lodestone, according to Gilbert's *De Magnete* (1600). The needle is parallel to the surface at the equator, perpendicular to the surface at the poles (which in this figure are located at the ends of the line marked "Axis").

off the coast of Antarctica or among the Arctic Islands of Canada's far north.

It was also known, long before Gilbert, that a compass needle often deviates from true north (determined from the stars) by several degrees. This discrepancy, today called "declination," varies substantially around the globe and has to be taken into account by navigators when reading their compasses. Declination results in large part from the fact that the earth's magnetic poles do not coincide with its geographic poles (Figure 3.3).

Declination of the compass from true north may actually have helped Columbus "discover" America in 1492. Like most captains of his time, Columbus relied heavily on his mariner's compass for navigation. His iron compass needles had coercivities so low that they had to be remagnetized frequently with a lodestone that he "guarded with his life," according to Samuel Eliot Morison, foremost chronicler of Columbus's voyages. Throughout the mid-Atlantic, Columbus set his course for due west. Because his compass needle was actually pointing slightly west of true north, however, his fleet was slowly trending southward and closer to

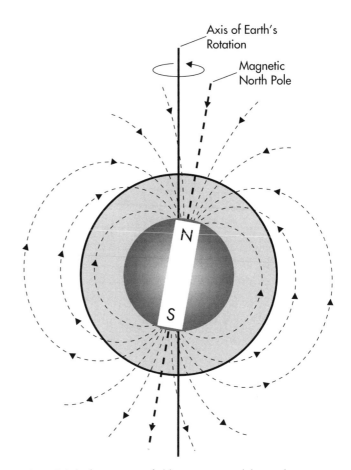

Figure 3.3 In the magnetic field pattern around the earth, use of a compass needle and dip needle to measure the field direction gives the approximate direction of north and approximate latitude, respectively. These measurements are not exact because the magnetic poles and geographic poles do not coincide.

the Bahamas. Landfall there came just in time, because the impatient crew, near mutiny, had been promised that the expedition would turn back soon if no land were sighted. If Columbus's course had really been true west, as he believed, it would have carried him north of the Bahamas (to the coast of Florida, several days' farther), and he might have been forced to return to Spain before landfall.

Later explorers, guided by their compasses but systematically checking (and correcting) for declination if clear skies allowed, soon crossed the much wider Pacific Ocean and circumnavigated the globe. Among them was Queen Elizabeth's favorite and William Gilbert's colleague, Sir Francis Drake. In *De Magnete,* Gilbert cites Drake as assuring him that the compass needle continues to point north even in the southern hemisphere.

Two centuries after Drake's voyages, another famed English explorer—Captain James Cook—embarked with scientists of the Royal Society for the first scientific expedition to the Pacific. After observing the transit of the planet Venus across the sun from the isle of Tahiti, Cook sailed south to survey the shores of New Zealand and Australia. In June 1770, while charting the eastern coast of Australia, he noted the unusual behavior of his compass needle, which he ascribed in his journal to "Iron Ore or other Magnetical matter lodged in the earth." The offshore island that Cook thought was attracting his compass he named Magnetic Island, which today is the site of Australia's Magnetic Island National Park. No magnetic anomaly has since been recorded near there, and historians suspect that some rearrangement of cannons or other magnetic material on the ship may have skewed Cook's compass that day. Nevertheless, the name has stuck, and present-day tourist promoters tout "the magnetic attraction" of this popular tropical island.

Another British explorer, James Clark Ross, located earth's magnetic north pole off one of the islands of northern Canada in 1831. He enjoyed his resulting celebrity and sailed for Antarctica ten years later in an attempt to duplicate his feat in the southern hemisphere. He got within about a hundred miles, reaching a dip angle of nearly 89 degrees, but his passage to the magnetic pole was blocked by ice. He wrote, "few can understand the deep feelings of regret with which I felt myself compelled to abandon the perhaps too ambitious hope I had so long cherished of being permitted to plant the flag of my country on both the magnetic poles of our globe." A near miss, but Ross Ice Shelf, Ross Island, and Ross Sea were all named after him as a consolation prize.

Although William Gilbert was successful in convincing the world that the earth itself was a magnet, he was less successful in convincing it how to label the ends of the compass needle. His experiments had clearly shown that when two magnets interacted, like poles repelled and unlike poles attracted (fact 2). It was obvious, he argued, that it was therefore the south pole of the compass needle that pointed north, and the north pole that pointed south! Everyone who had written about the compass before him was in error, "so ill-cultivated is the whole philosophy of the magnet still, even as regards its elementary principles." Gilbert's logic was impeccable, but people just didn't like labeling the poles of compass needles that way. In common usage, the poles that point north are called north poles (north-seeking) and the poles that point south are called south poles. This means, of course, that the earth's "north" magnetic pole is in the south, and vice versa, but that doesn't seem to bother most people today. (For some exceptions, see Chapter 13.)

///// **Reading the Rocks**

For many centuries now, sailors have used the earth's magnetic field as revealed by compass needles to guide their course (after correction for declination). In the latter half of the twentieth century, geologists have learned how to use the earth's magnetic field, as revealed by rocks serving as "fossil compasses," to guide them to a greatly improved understanding of the earth's crust, including the motion of continents and the widening of oceans.

Only a small fraction of the earth's rocks contain enough magnetite to act as lodestones. However, many rocks contain small amounts of magnetic material, enough to allow the direction of the rock's magnetization to be measured with sensitive laboratory equipment. The field of *paleomagnetism* is based on the concept that the direction of magnetization observed in each rock records the direction of the earth's magnetic field at the place and time the rock was formed. Rocks with sufficient coercivity will retain this memory (a long-term "magnetic recording") despite later changes in the direction of the earth's field.

Geologists have also developed various techniques to estimate the ages of rocks—how many millions of years in the past they were formed, either by the cooling of molten lava or by settling out as sedimentation at the bottom of some ancient sea. Researchers combined these data to learn about the history of the earth's magnetic field—but they learned a lot more! One geologist was so impressed with how much was learned about the earth's history by reading the fossil magnetism stored in rocks that he dubbed these rock-reading scientists "paleomagicians."

Their first studies, on rocks only a few million years old, showed that the earth's magnetic poles, when the rocks were formed, were near their present positions. But sometimes the poles had reversed! In other words, it appeared that the north pole of the global magnet became the south pole, and vice versa, at certain times in earth history. Records of field reversals in rocks from various parts of the world were consistent, showing convincingly that the earth's magnetic field had reversed many times in the past, about once every half-million years on average.

The magnetic readings of rocks became more mysterious as geologists studied older and older rocks. At first, results suggested that the magnetic poles had wandered large distances, and had not always been near the geographic poles. Even more disturbing, rocks of the same age from different continents indicated different positions of the magnetic poles. After some years of confusion, it was finally realized that the magnetic poles, except during the brief periods in which they were in the process of reversal, had stayed near where they are today all along. It's the continents that had moved!

The "fossil compasses" provided strong support for the theory of continental drift, previously very controversial. Today continental drift is a universally accepted part of plate tectonics, the general model geologists use to explain earthquakes, mountain formation, volcanoes, and other geological activity. The model has also been used to map, for example, the positions of the continents around the Atlantic at various times in the past (Figure 3.4). The fitting of the coastlines of Africa and South America, like pieces of a jigsaw puzzle, was one of the early clues that the continents are mobile.

Further evidence supporting continental drift came from measurements of anomalies in the earth's magnetic field underneath the Atlantic Ocean. As seen in Figure 3.4, Europe and North America are gradually drifting apart, and the Atlantic is progressively widening as new sea floor is created along a mid-oceanic ridge. Lava flows up through the mid-oceanic ridge and becomes magnetized in the direction of the earth's magnetic field at the time it solidifies. Half of the ocean floor moves east, half moves west as the new rock pushes the old sea floor away from the ridge, and the distance from the mid-oceanic ridge becomes a measure of the age of the rocks. Over millions of years, reversals in the earth's field therefore leave parallel bands of oppositely magnetized rocks on the ocean floor (Figure 3.5). The spreading ocean floor serves as a *very* slowly moving tape recorder, encoding the history of the earth's magnetic field.

The rocks on the sea floor, magnetized either parallel to or opposite to the present earth's field, produce slight increases or decreases in the local field. These magnetic anomalies have

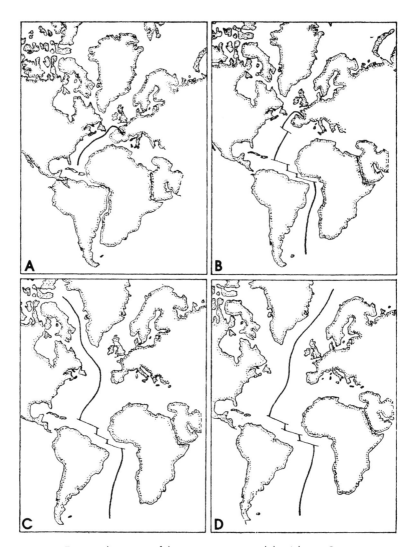

Figure 3.4 Estimated position of the continents around the Atlantic Ocean at various times in the past: (A) early Jurassic, 170 million years ago; (B) mid-Cretaceous, 100 million years ago; (C) early Paleocene, 65 million years ago; and (D) Eocene-Oligocene, 40 million years ago. The black line between the landmasses indicates the position of the mid-Atlantic ridge (where sea-floor spreading takes place) at each stage.

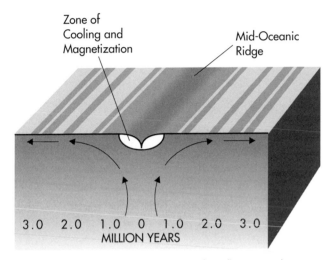

Figure 3.5 A schematic representation of sea-floor spreading. Molten rock is not magnetized, but as the magma cools it takes on the magnetization of the field in which it solidifies. Lava flowing up through the mid-oceanic ridge therefore leaves a record in the sea floor of the direction of the earth's magnetic field at the time it solidified. The pattern of magnetic reversals across the ocean floor, measured with magnetic anomaly detectors (MADs), thus records the history of the reversals of the earth's field over time; shown here is a record of the past 3 million years, corresponding to a distance of about 50 miles. (Black bands are magnetized in the direction of today's field, light bands in the opposite direction.)

been measured by very sensitive magnetic-anomaly detectors—the MADs mentioned in the previous chapter—towed across the ocean by research ships. Put another way, the MADs serve as "pickup heads" to read the "tape recorder" on the bottom of the Atlantic. By comparing the various field reversals observed on the ocean floor with data from lavas on land, a time scale has been established for the widening of the Atlantic. Analysis reveals that Europe and North America are separating at a rate of over one-half inch per year. (This may eventually lead to an increase in trans-Atlantic airfares.)

///// Undercurrents

Gilbert showed that the earth is a great magnet—a magnus magnes—and rocks tell us it has been a great magnet (although occasionally reversing polarity) for billions of years, perhaps for all of its 4.5 billion years of life. What is the origin of its magnetism? In 1905, Einstein called this one of the most important unsolved problems in physics. Scientists now feel this longstanding problem is partly solved, although still incompletely understood.

Gilbert thought that the earth was a permanent magnet—a giant lodestone. But now that we have accurate measurements of the earth's field over several hundred years, we know that the field is constantly changing, both in magnitude and direction. For example, the declination (deviation from true north) in London was 11 degrees east in 1580, but it had drifted to 24 degrees west by 1800 and back to 6 degrees west by 1963. The inclination, or dip angle, also changed by several degrees over this time period. These data indicate a wandering of the magnetic poles by several hundred miles. The magnitude of the earth's field has also changed, decreasing by several percent since 1580. In addition, we know from reading the rocks that, over a much longer time scale, the earth's field has actually reversed many times. All these changes, slow as they may seem to us, would not be expected if the earth were a permanent magnet.

We also know more about the interior of the earth than Gilbert did. In particular, we know that the temperatures in most of the earth's interior are far above the temperatures at which lodestones, iron, and all other magnetic materials lose their magnetism. We have to look for some other explanation of the earth's magnetic field. (Each material loses its magnetism above a characteristic temperature, named the *Curie temperature* after the French physicist Pierre Curie, who studied the phenomenon in considerable detail. The fact that sufficient heating destroyed magnetism had been recorded much earlier by Gilbert.)

The radius of the earth is about 4,000 miles. Seismic evidence shows that, about halfway down, the earth is no longer solid. The liquid core of the earth is metallic, and electric currents circulating in this conducting liquid—billions of amperes of current—are the source of the earth's field. The earth is not a permanent magnet—it's an electromagnet!

Recall from Chapter 1 that electric currents produce magnetic fields (fact 6) and that changing magnetic fields induce electric currents (fact 8). Applying this knowledge to the complex problem of a rotating conducting liquid with large variations in temperature and density has produced a "self-excited dynamo" theory that can explain most of what we know about the earth's magnetic field. The theory assumes the existence of an initial nonzero magnetic field, perhaps from the sun. Motion of the conducting liquid core in this field yields electric currents (fact 8), and the electric currents produce magnetic fields (fact 6). A continuous source of energy is needed to drive this bootstrap process, energy that probably comes from heat released by the gradual solidification of some of the liquid core as the earth cools. If the dynamo theory is correct, the magnetic fields within the liquid core of the earth are larger and more complex than the fields we measure at the surface. Instabilities in the patterns of fluid flow and electric currents—in the earth's deep undercurrents—produce the changes over time in the direction and magnitude of the surface field that we observe, including the reversals.

Magnetized rocks in the earth's crust cause small local variations in the earth's field, including those that have revealed the widening of the Atlantic. The motion of charged atoms and electrons miles above our heads—electric currents in the ionosphere and magnetosphere—also contribute to minor variations. (These fields have a small daily fluctuation and also vary with changes in solar activity.) But most of the earth's magnetic field, the field that moved Einstein's compass needle and guided Columbus's ships, comes from electric currents flowing in molten metal over 2,000 miles beneath our feet. "Something deeply hidden" indeed!

///// Cosmic Currents

There are many large magnets in the universe other than our own earth, we now realize. NASA's interplanetary probes have gathered data on magnetic fields in our solar system, and we can also measure more distant magnetic fields, even from faraway stars and galaxies, by their effects on light and radio waves that reach the earth.

In our solar system, most planets have been found to have magnetic fields of their own, probably generated by internal dynamos. Jupiter's surface field is about four gauss, ten times larger than the earth's. Mercury's is a hundred times smaller than the earth's. "Farout" planets Uranus and Neptune, we learned from the recent flyby of *Voyager 2*, both have large magnetic fields, but their magnetic poles are tilted far away from their geographic poles. Our neighboring planets Mars and Venus, on the other hand, have little or no magnetic field. Further study of these wide variations among the planets may help us better understand the earth's internal dynamo.

Our moon also has no magnetic field of its own. Considering the connections noted in the previous chapter between magnets and romance, it is particularly disappointing to learn that neither the moon, subject of countless love songs, nor Venus, named for the goddess of love, is magnetic. Moon rocks brought back by Apollo astronauts were magnetized, however, suggesting that the moon may have had a magnetic field in its distant past.

Perhaps Venus did as well. (Experts believe that the very slow rotation rate of Venus—one rotation every 243 earth days—may explain its lack of a magnetic field. Love may "make the *world* go round," but it is apparently rather ineffective on Venus.)

In 1991, the *Galileo* spacecraft got close enough to Gaspra, an asteroid about 8 miles across, to determine that it had surface magnetic fields about equal to earth's. Gaspra's too small to have an internal dynamo, so it must be a permanent magnet—a huge, flying lodestone. Scientists are not sure where or when it got magnetized. Perhaps it is a fragment of a former planet and was magnetized there. Or perhaps it was magnetized in the early history of the solar system, when the interplanetary magnetic field may have been much larger than it is today. Another mystery.

Many stars are known to have intense magnetic fields. Some so-called magnetic stars have fields over 10,000 gauss, white dwarfs have fields up to 100 million gauss, and neutron stars have fields of a trillion gauss! Interstellar magnetic fields are rather weak, only a few millionths of a gauss, produced by dynamo action in the interstellar gas, a thin conducting gas carrying feeble electric currents over large distances. A galaxy is a weak, but very large, electromagnet, about a quintillion (million trillion) miles across. That's a real *magnus* magnes. (In the next chapter, I'll introduce simpler ways of representing very large and very small numbers.)

Our own star, the sun, has an average surface magnetic field of about one gauss, the direction reversing every eleven years. The dynamo generating the sun's field is apparently much more turbulent than the earth's. Dynamic surface features like sunspots exhibit local magnetic fields of a thousand gauss or more, evidence of east-west fields within the sun much higher than the average north-south field seen on the surface. Related solar flares emit clouds of charged particles that reach the earth in two or three days and are deflected by the earth's magnetic field to the polar regions. There they produce intense auroral displays, called "northern lights" in my hemisphere.

Solar flares also cause changes in the earth's magnetic field, known as "magnetic storms." Although the field changes in magnetic storms are only a small percentage of the earth's average field, they can occur rapidly and over large distances. These field changes induce electric fields and currents (fact 8 again) and can lead to significant current surges in electric power lines. Intense magnetic storms in 1989 were blamed for the failure of two transformers in New Jersey, three capacitor banks in Virginia, and a five-hour blackout of the entire province of Quebec!

Solar flares, essentially magnetic "burps" of the sun's dynamo, clearly can have substantial effects here on earth. One study even reported a correlation between magnetic storms and admissions to psychiatric hospitals, as though instabilities in the sun's dynamo induced, a day or so later, magnetic "brainstorms" on the earth. Skepticism about a cause-and-effect relationship here is warranted, but the brain is a complex electromagnetic system we don't yet quite understand. If the varying magnetic fields in magnetic storms can blow out transformers and capacitor banks, maybe they can induce a few electrical disturbances in our heads as well.

///// Biocompasses

The fields of biology and magnetism became inseparably joined in 1975, when microbiologist Richard Blakemore demonstrated that some aquatic bacteria *(Aquaspirillum magnetotacticum)* navigate with the help of internal magnetic compasses. Originally discovered in mud collected from the marshes of Cape Cod, Massachusetts, and the bottom of nearby Buzzards Bay, these bacteria swam toward the south pole of a bar magnet and away from its north pole!

Electron microscopy revealed that inside each bacterium was

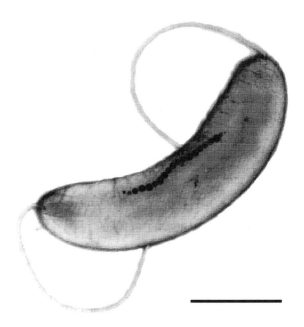

Figure 3.6 An internal compass: electron micrograph of a bacterium with magnetite particles inside. The earth's magnetic field exerts a torque on the chain of magnetite particles, aligning the bacterium with the field. The bar corresponds to a length of one micrometer (one millionth of a meter).

a chain of tiny crystals of magnetite—tiny lodestones—each only a few millionths of an inch in diameter (Figure 3.6). These chains are permanent magnets, compass needles with coercivities of about 300 gauss (like lodestones) and north poles at their front ends. (I was tempted to say at their heads, but, as you can see from the picture, that doesn't seem appropriate.) These bacteria therefore swim north, and swim *down* in the mud because of the inclination or dip of the earth's magnetic field.

Note that, like ordinary compasses, the bacteria are not being pulled north or south. The earth's magnetic field exerts equal and opposite forces on the ends of the chain of magnetite particles, producing a torque that rotates it until it lines up with the field (Figure 3.1). The bacteria supply their own power (rotating their flagella) to move along the direction in which they have been pointed by the earth's field.

What advantage did these creatures derive from developing

an internal compass? Moving north has no advantage to them, but moving *down* takes them toward the bottom of the muck in which they live, away from the oxygen-rich environment at the top, which is toxic to them. Bacteria living in muds in the southern hemisphere, where the dip angle is reversed, are found to have south magnetic poles at their front ends, directing them south and down into the sediment, away from that unhealthful (for them) oxygen. Near the magnetic equator, where the dip angle is near zero, north-seeking and south-seeking bacteria are equally prevalent. Here both types at least are guided horizontally, reducing harmful upward movement.

In the laboratory, you can change north-seeking bacteria into south-seeking bacteria simply by switching on a large reverse field, one that's larger than the coercivity of the magnetite, thereby reversing the poles on their compass needles. The bacteria don't complain—they just aim for the opposite pole.

Since Blakemore's studies, many other living organisms, including humans, have been found to contain small bits of magnetite and other magnetic compounds of iron somewhere in their anatomy. In some cases, like the bacteria, these compounds may be used by the organism as biocompasses for navigation. Some experiments suggest, for example, that homing pigeons and honeybees may use the earth's magnetic field to assist in navigation (Chapter 15).

In other cases, the iron compounds may simply be by-products of metabolism. For example, a bacterium known as GS-15 generates magnetite, but in a form not useful as a biocompass. Recent studies suggest that fossilized GS-15 and other magnetic bacteria, some containing biocompasses, may have contributed a large fraction of the magnetite found in sedimentary rocks. Millions of years after the rocks were formed, geologists will use this magnetite (along with magnetite in igneous rocks formed from cooling lava) as fossil compasses to study the history of the earth's magnetic field, continents, and oceans.

Life evolved on earth in the presence of a magnetic field, and at least one form of life, Blakemore's bacteria, evolved to make good use of it. Humans eventually learned to make use of it, too, with the help of lodestones and compass needles, to cross the oceans. But until the nineteenth century, they had done little to change the magnetic fields in which they lived. Since then, however, development of electromagnetic technology has accelerated steadily—producing many wonderful devices to use and to enjoy, but also altering the electric and magnetic fields that permeate our bodies daily. We will come to the complex and controversial topic of magnetic fields and health later, but before that there is much to learn about our successes in improving on nature's magnets and some of the many ways in which we use magnets in our technology today.

///// SUPERMAGNETS

4

///// A New Science

Today we have permanent magnets hundreds of times stronger than nature's lodestones, and superconducting electromagnets far stronger than the best permanent magnets. Although the science and technology of magnets have a very long history, the pace of improvement in magnet materials has quickened remarkably in recent years with the maturing of a newly identified field of applied science: *materials science.*

I now teach in a department of materials science and engineering. When I went to college, there were no departments with that name, even though metallurgists, ceramists, and scientists of other disciplines were already applying the concepts of physics and chemistry to better understand the relationships between the manufacture, the microstructure (how the atoms are arranged), and the properties of engineering materials. With the expansion of this approach to include the study of polymers, semiconductors, and other solids of engineering significance, the field of materials science emerged as a "new" academic discipline.

In dozens of universities around the world, metallurgy departments were renamed materials departments. The Materials Research Society, founded in 1973, is today one of the world's fastest-growing professional societies. To meet this competition, established societies changed their names. The Metallurgical Society became The Minerals, Metals, and Materials Society, and the American Society for Metals, postal address Metals Park, Ohio, became ASM International, at Materials Park, Ohio. This is probably the first time in history that the development of a new academic discipline required the renaming of a post office! Perhaps most important, government funding agencies today explicitly recognize "advanced materials" as a high priority for funding support. Materials are definitely "in." Why?

As Madonna sang in what became her trademark song, "We are living in a material world (and I am a material girl)." In our material world, every device we use is made, of course, from materials—metals, semiconductors, ceramics, plastics, composites—and the performance of all our devices is limited by the properties of materials. The task of materials scientists is to develop new and improved materials, or new and improved processes to manufacture materials, and thereby broaden the options available to technology. The payoff from investments in this area of applied science in recent decades has been impressive. Improved semiconductors, superconductors, optical materials, structural materials, and magnetic materials have led to remarkable advances in technology, only a few of which we can touch on in this book. Materials science has a good track record, and U.S. support of research and development in this area has become an important part of our national strategy for industrial competitiveness.

The building blocks of the materials scientist are the hundred-odd chemical elements of the periodic table. Like musicians with their notes, and painters with their colors, materials scientists experiment with new combinations and arrangements of this finite set of chemical elements. As a necessary foundation for materials science, we first had to separate and identify those elements—a task that took many centuries.

///// The Elements of Things

To Aristotle, there were only four basic elements: earth, air, fire, and water. Nevertheless, seven substances recognized today as elements were also known in his day—gold, silver, copper, iron, lead, tin, and mercury—because they can be found in nature, or easily produced, in relatively pure form. (Pure iron was found mostly in meteorites. Gilbert thought that this iron that "rained" was formed "in the air, in the uppermost clouds from the earth's vapors.") In the Middle Ages, alchemists related these seven metals to the seven heavenly bodies seen to move among the fixed stars. They coupled gold with the sun and silver with the moon. Venus represented copper and woman, softness and beauty; Mars represented iron and man, strength and warfare. Pretty sexist, but we still use the alchemist's symbols for iron and copper to represent male and female.

In 1661, 2,000 years after Aristotle, Robert Boyle finally articulated our modern concept of chemical elements. Chemical techniques were developed to separate compounds into their component elements, and by 1800 sixteen other elements had been added to the original seven. By 1900, most of the elements known today had been identified.

Iron, number 26 in the periodic table and the most important element in the world of magnets, occupies a very special place among the elements. Its nucleus is the most energetically stable of all. Atomic nuclei contain two types of "nucleons"—positively charged protons and uncharged neutrons—and, of all the elements, the nucleus of iron has the lowest energy per nucleon.

Lighter elements can release energy by *nuclear fusion* to form heavier elements. Energy production in our sun results primarily from the fusion of two hydrogen nuclei (atomic number 1) to form helium (atomic number 2). Other elements—lithium (3), beryllium (4), boron (5), carbon (6), etc.—can be formed by fusion of one or more lighter elements in hotter stars. But energy-releasing fusion reactions can only produce elements up to iron (26).

The creation of elements heavier than iron instead *consumes* energy. These elements are believed to have formed in supernova explosions. (As has often been said, we and everything else on the earth are "stardust.") Rather than giving up energy upon fusion, heavy elements can release energy by splitting into two lighter elements—by *nuclear fission*. The most familiar example is the fission of uranium, element number 92 (see Chapter 11).

So light elements can fuse and move *up* in atomic number, releasing energy in the process, and heavy elements can split and move *down* in atomic number, also releasing energy. In a sense, they're all aiming to be as close as possible to iron, the element with the most stable nucleus—the lowest nuclear energy per nucleon—of all.

One result of the nuclear stability of iron is that there is more iron in the universe, and in the earth, than any other metal. Most of the earth's core is iron (although it is too hot to be magnetic), and, as you may recall, iron is the fourth most common element in the earth's crust. Although aluminum and silicon are more common in the crust, they're harder to separate from their oxides. For these reasons, iron and steel have a dominant position in our technology quite independent of their importance as magnets. Believe it or not, the main reason that refrigerators are made of steel is not because magnets stick to it. Steel is strong, formable, and cheap—a tough combination to beat. Although there are many technological devices in which steel is used because of its magnetic properties, in most applications its high strength and low price make steel the material of choice.

Are we stuck with just the hundred-odd elements we know

today, or will scientists of the future have other colors on their palette, other notes on their keyboard? I was recently teaching a materials course to students in aeronautical and astronautical engineering, and one student showed me the "Technical Manual" purportedly describing the construction of the starship *Enterprise* of *Star Trek: The Next Generation.* The manual indicates that alloys of "tritanium," "duranium," and other seemingly new elements were the main structural materials. I argued that these must just be new names for our present-day elements, because no new stable elements are likely to be found. The student was polite, but he probably just thought that the old fogey was being characteristically conservative. (I should perhaps mention that nuclear physicists calculate that some "superheavy" elements, with atomic numbers and weights beyond the elements currently known, may be stable if they could somehow be formed. But superheavy elements seem unlikely to be useful as structural materials for starships.)

Centuries from now, magnets and magnetic materials will be much improved over those we have today, but they probably won't be made from new elements. They'll be made from new combinations and arrangements of the same elements we know now. And so will starships.

///// Improving on Lodestones

Gilbert's *De Magnete* of 1600 reported that steel retains its magnetization better than iron; today we'd say it has a higher coercivity. But its coercivity was still so much lower than that of nature's magnet, the lodestone, that lodestones remained the best permanent magnets for many centuries despite having much lower saturation magnetization than steel.

Iron was usually hardened into steel by plunging hot iron into various quenching liquids. (Liquids recommended by an anonymous expert in 1532 included radish juice, dragon's blood, "men's pisse and the distilde water of wormes." Evidently there was still lots of magic in steelmaking in those days.) From ancient times through the Middle Ages, people knew how to make steel but they didn't quite know what steel was. Aristotle thought that steel was simply a pure form of iron; we now know it's just the reverse. It wasn't until 1781 that Swedish chemist Torbern Bergman showed that steel was simply iron containing about one percent carbon.

Knowing how steel differed chemically from iron made it a lot easier to produce it reliably. Improved carbon steels, along with improvements in magnetizing techniques and mechanical design (among them, the now-familiar horseshoe shape), finally produced steel magnets more powerful than lodestones. Carbon steels remained the world's strongest permanent magnets until about 1880, when additions of tungsten and other elements produced even better magnets.

Among those interested in magnets a century ago was Marie Curie, later a two-time winner of the Nobel Prize (in 1903 for physics, and in 1911 for chemistry). As mentioned in the previous chapter, her husband, Pierre Curie, also did research on magnets. He shared the 1903 prize with her for their work on radioactivity, making them the only husband-wife team to win the Nobel Prize. In 1897, Madame Curie studied various alloy steels and found that steels containing tungsten, chromium, or molybdenum all had coercivities near 80 gauss, higher than the best carbon steels. Tungsten steels were the magnets of choice until World War I, when a shortage of tungsten brought chromium steels to the fore. In 1920, Japanese scientists found that adding cobalt to chromium steels more than doubled coercivities. The strongest magnets then contained four elements—iron, carbon, chromium, and cobalt.

Many materials today are composed of several chemical elements. Some alloys for jet engines, for example, contain up to twelve elements, each added for a specific purpose. These are said to contain "everything but the kitchen zinc."

As more and more different magnet alloys became available to engineers, the need arose for a single measure of quality with which to compare them. Steel A may have a higher coercivity (be more "permanent") than steel B, but a lower saturation magnetization (be less "magnetic"). Which steel is better? The *maximum energy product*, a quantity that depends on both coercivity and saturation magnetization, became accepted as the best single measure of the quality of a permanent magnet. It is commonly measured in units of gauss-oersted (GOe). (Hans Christian Oersted was a Danish physicist who in 1820 was the first to discover that an electric current created a magnetic field. This was a turning point in the history of science and technology, but it was not a fairy tale, so in Denmark Oersted is often referred to as "the *other* Hans Christian.")

The cobalt-chromium steels, the strongest magnets available in the 1920s, had a maximum energy product of one million

PERMANENT-MAGNET TERMINOLOGY

saturation magnetization   The magnetic field produced by a magnet when all its atomic magnets are aligned; a measure of the ultimate lifting strength of a magnet.

coercivity   The strength of the reverse magnetic field required to demagnetize a magnet after it has been magnetized; a measure of the "permanence" of a permanent magnet.

maximum energy product   A common quality index for permanent magnets that depends on both saturation magnetization and coercivity; the higher the energy product, the smaller the size of the magnet needed for a specific application.

GOe (one megagauss-oersted—l MGOe), four times stronger than carbon steels and almost ten times stronger than most lodestones.

Steel magnets were topped in the 1930s by *alnico* magnets. (The Scottie dogs I played with as a child were mounted on alnico magnets.) As their name implies, alnicos contain aluminum, nickel, and cobalt, but they are still more than 50 percent iron. Additional alloying elements and variations in manufacture led to a range of alnicos, some with energy products as high as 8 MGOe. Alnico magnets remain in use today but are steadily losing market share to two newer classes of permanent magnets—ferrites and rare earths.

The term *ferrite* is commonly applied to a wide range of compounds of iron oxide with other oxides. Chemically, ferrites are more like lodestones than like steels or alnicos. Steels and alnicos are metals—electrical conductors—while ferrites and lodestones are oxides—electrical insulators. (This difference in electrical properties is important only in applications in which the magnet is exposed to varying magnetic fields. Varying magnetic fields will, via fact 8, induce electric currents in metallic magnets but not in insulating ferrites.) First introduced in the 1950s, ferrites had surpassed alnicos in annual tonnage within about ten years. Today they account for over 90 percent of the permanent-magnet market (by weight), despite having energy products less than the alnicos. Why did inferior magnets enjoy such market success? They do have higher coercivities than the alnicos, which is useful in some applications, but their major advantage is that they are *cheap*.

One thing I learned working at General Electric that they didn't teach me at Harvard is the importance of price in materials selection. For many consumer applications (including refrigerator magnets), a better figure of merit than energy product is *energy product per dollar*. There, ferrites win hands down.

For applications where price is secondary to energy product

or coercivity, the clear winners are the newest and currently most exciting class of permanent magnets—the "rare-earth" magnets.

The rare-earth metals include the lanthanides (elements 57 to 71), yttrium (39), and scandium (21). The two that have become most important for producing permanent magnets are neodymium (60) and samarium (62). Commonly occurring together in ores as mixed oxides, and with similar chemical properties, the individual rare-earth metals were difficult to separate and were not available in high purity or in large quantities prior to the 1940s.

During World War II, workers on the atomic bomb project learned that nuclei of the rare earths were prevalent among the products of uranium fission, and development of practical means of chemically separating these elements became a high priority. Techniques successfully developed at Iowa State University and at Oak Ridge, Tennessee, made it possible, for the first time in history, to produce large quantities of high-purity rare-earth metals. After the war, researchers around the world suddenly had the opportunity to obtain sixteen newly available elements and to study their physical properties. By the 1960s several groups had progressed to studying the magnetic properties of compounds formed between the rare earths and other elements.

The first commercial rare-earth magnets, introduced in 1970, were based on a compound of samarium (a relatively rare rare-earth metal) and cobalt, and were developed at the General Electric Research Laboratory in Schenectady, New York. I had the good fortune to be working at GE at that time. Although my role in this magnet development was only a peripheral one, I'm glad that I was there to see, firsthand, the birth of a revolutionary new class of permanent magnets. Within a few years, magnets with coercivities over 20,000 gauss, and magnets with energy products over 25 MGOe, reached the market. Unfortunately, GE left this game after the first inning. Shortly after the first cobalt-samarium magnets were produced, GE sold its magnetic materials business, including all the pertinent patents, to Hitachi. GE management saw this exciting new materials development as an opportunity to sell off a business that had been relatively unprofitable in the past, rather than as an opportunity for future profits. Technological leadership in rare-earth magnets moved to Japan.

Political upheavals in Zaire, source of much of the world's cobalt, led to cost and supply problems with these new magnets in the late 1970s. A worldwide search for cobalt-free permanent magnets began, and in 1983 two companies announced their development of magnets based on iron (the old reliable), boron (a probably serendipitous addition), and neodymium (a rare-earth metal considerably less rare than samarium). Some "neo" magnets have since exceeded 50 MGOe in energy product—fifty times stronger than the cobalt-chromium steels that were the strongest magnets available when I was born, and hundreds of times stronger than lodestones. From this perspective, they are truly "supermagnets," the latest of a long line of remarkable improvements in permanent-magnet materials made in this century (Figure 4.1).

As energy product increased, the amount of magnet required for a specific application decreased, as did the size and weight of many devices employing permanent magnets, including, as we'll see in the next chapter, motors, speakers, and telephone receivers. In many cases, the great improvements in magnet properties allowed the creation of new or newly designed devices that would have been impossible or impractical with earlier magnets.

Much of the increase in energy product over this century resulted from increases in coercivities, which were less than 100 gauss for early alloy steels and are over 10,000 gauss for most rare-earth magnets. This has influenced the optimum shapes of magnets. Low-coercivity magnets had to be long, to minimize the demagnetizing effects of the magnetic poles at each end of the magnet. Thus steel magnets, with both low energy products and low coercivities, had to be both large and long. This require-

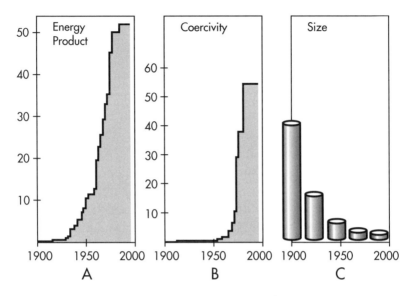

Figure 4.1 The remarkable increases in (A) energy product (in megagauss-oersted) and (B) coercivity (in kilogauss) of permanent-magnet materials in this century. As magnets became more powerful, the size and length of a magnet required for a specific application (C) decreased.

ment led naturally to the horseshoe-shaped magnet, which is large and long but allows both poles to contact the item to be lifted. Modern magnets, with high coercivities, are seldom horseshoe-shaped. As shown in Figure 4.1, a small disk of rare-earth magnet can now do the same job that formerly required a long and large steel magnet. Yet the horseshoe image of magnets persists, even though the steel magnets most suited to that shape (Figure 2.3) became obsolete over a half-century ago. Old images die hard.

What do these increased energy products mean for our daily lives? A small piece of rare-earth magnet can now do the job that, in my youth, required a steel magnet fifty times bigger. In 1993, nearly 3,000 tons of iron-neodymium-boron "supermagnets" were produced worldwide and incorporated into motors and other devices, but since most magnets remain hidden, few people are aware of their significance in today's technology. Rare-earth magnets provide smaller and lighter starter motors in General Motors cars, and perform key functions in computers and other high-tech devices. The most familiar examples are perhaps the compact earphones for Walkman-type cassette players and radios, which are made possible by rare-earth magnets. Very few of the teenagers listening to the latest rock or rap through their earphones today realize the debt they owe to improved permanent magnets.

Although the strongest magnets are solid metal or, in the case of ferrites, solid ceramic, magnets may also be made of powdered metal or ceramic imbedded in a matrix of rubber, plastic, or an easily melted metal. The resulting "bonded" magnets are weakened by dilution in a nonmagnetic matrix, but they have the advantage that they can easily be formed into complex shapes with accurate control of dimensions. They are also mechanically sturdy, which cannot be said of solid metallic or ceramic magnets, which are often quite brittle. Finally, they are less prone to corrosion, which can be a problem with some rare-earth magnets, particularly "neo" magnets.

The two companies that led in the early development of "neo" magnets were Sumitomo and General Motors (although GE was no longer a player, GM kept the U.S. in the game). Japan still leads the world in the production of rare-earth magnets, but second place now belongs to China. The site of a high percentage of the world's rare-earth ores, China has learned to produce high-quality rare-earth magnets at moderate cost, has negotiated patent agreements with Sumitomo, and in 1995 purchased General Motors Magnequench, the division that produces neo magnets. Many centuries ago, the Chinese led the world in recognizing the north-south directivity of magnets. Today they are making a strong effort to achieve world leadership in magnets once again.

///// Better and Bitter Electromagnets

However wonderful permanent magnets may be, they have two distinct limitations. First, the amount of magnetic field they can produce is limited by their saturation magnetization, currently limited to about 15 kilogauss (15,000 gauss). To judge from current knowledge, 20 kilogauss is an optimistic upper limit for possible future development. For higher magnetic fields, you need an electromagnet.

Second, permanent magnets are hard to turn off! If James Bond had used a permanent magnet instead of an electromagnet, he could have suspended Jaws over the shark pool, but he couldn't have dropped him into it. In our AC electromagnetic world, many applications actually require magnetic fields that reverse many times a second. For changing magnetic fields, you need an electromagnet. We'll come to AC electromagnets in Chapter 8, but for now consider the other important use of electromagnets— the production of very high magnetic fields.

The only way to get to fields much higher than 20 kilogauss is to use lots of electric current. You can do this with superconductors, the subject of the next chapter, or you can do it with normal metals and "brute force." It was Francis Bitter of MIT who led the way toward producing high-field electromagnets with the latter approach. Because high currents would produce a great deal of heat in the copper coil, Bitter knew that he needed two things—lots of electric power and lots of cooling water. Using specially designed conductors, 1.7 *million* watts of electrical power, and 800 gallons of water per minute, in 1936 he was the first to generate a continuous magnetic field of 100 kilogauss. The conductors in his "Bitter magnets" were not wires but perforated disks, designed not only for efficient cooling (water flows through the perforations) but also for standing up to the large forces associated with high magnetic fields.

Many properties of materials change in the presence of large

Figure 4.2 A Bitter electromagnet at the Francis Bitter National Magnet Laboratory at MIT. In the foreground are water-cooled cables that supply current up to 39,000 amperes. Purified cooling water flows through the magnet at the rate of 1,500 gallons per minute. This brute-force electromagnet requires 8 million watts of electrical power to reach a magnetic field of 230 kilogauss.

magnetic fields, and researchers use those measured changes to investigate the physics and chemistry of various materials. Bitter's pioneering magnets, and the recognized need for high magnetic fields for researchers throughout the country, led to the formation of the National Magnet Laboratory at MIT in 1960. After his death in 1967, the lab was renamed the Francis M. Bitter National Magnet Laboratory. Various improvements on his original Bitter magnets later produced continuous fields over 200 kilogauss, which required eight million watts of power (Figure 4.2).

The National Science Foundation decided in 1990 to move the

National Magnet Laboratory to Florida State University (working with the University of Florida and Los Alamos Scientific Laboratory). The new laboratory in Tallahassee was formally dedicated in October 1994. Why the move? Despite fifty years of leadership in electromagnet technology, and despite receiving superior reviews of their renewal proposal by scientific experts, MIT lost the NSF grant on what appear to be financial and political grounds. Florida offered to share the costs of the new laboratory, and Florida now has more clout in Washington than Massachusetts, a state that some feel already receives more than its share of federal research funding.

For scientists who wanted to do experiments in even higher fields, and who could design their experiments to make their measurements in a fraction of a second, several labs developed high-field *pulsed* magnets. Pulsed magnets pass huge currents, but heating is limited because the currents are very brief. Fields up to 500 kilogauss have been generated for a second or so, and fields up to a million gauss have been generated for tenths of a second. At these and higher fields, forces are so strong that the coils are often destroyed in one pulse! Designing coils strong enough to withstand large forces and still conductive enough to carry large currents has been a difficult materials challenge, since most high-strength materials, like steels, are poor conductors. However, a recently developed copper-niobium composite conductor has allowed repeated nondestructive pulses, nearly a tenth of a second in duration, up to 730 kilogauss, and nondestructive pulses of a million gauss or more are likely to be possible soon.

These already huge fields can be further increased by "implosion" techniques, in which coils are surrounded by high explosives that, when detonated, compress the magnetic fields up to 10 million gauss or more. These are spectacular fields, approaching those in white dwarf stars (and perhaps helping us understand atomic and nuclear processes in such stars), but the experiment is very noisy and very brief. The equipment self-destructs in less than a microsecond (a millionth of a second). These imploding electromagnets at least have provided the supporting funding agencies with, literally, a "big bang for the buck."

Most engineering applications, and even most scientific experiments, can't make use of magnetic fields—even very high magnetic fields—that last only a fraction of a second. So pulsed magnets, impressive as they are, are of limited usefulness. If all you need is a steady but modest field, you can use a permanent magnet or a small electromagnet coupled with iron to enhance the field. If your needs are not so modest, you can get steady fields up to 40 kilogauss or so with a large iron-core electromagnet, but such magnets are costly, heavy, and power-hungry. For steady fields up to about 200 kilogauss, you can visit the National Magnet Lab and use the Bitter magnets.

Until about 30 years ago, those were your only choices. You now have another: electromagnets that produce high, steady magnetic fields using only tiny amounts of electrical power. These magnets, now commonplace in laboratories and hospitals around the world, are based on a remarkable phenomenon discovered in 1911—superconductivity—and the discovery a half-century later of a new class of materials, known as high-field superconductors. We discuss those in the next chapter, but first a brief interlude to consider the size of things.

///// Microthings and Megathings

In this and the preceding chapters, we have encountered very large magnetic fields (in neutron stars and pulsed magnets) and very small magnetic fields (in interstellar space). Scientists don't like to write out the long strings of zeros needed to express very large or very small numbers, like 1,000,000,000,000 for a trillion, or 0.000001 for a millionth. We prefer either to use exponential or "scientific notation," like 10^{12} for trillion or 10^{-6} for millionth, or to substitute prefixes for numerals, like *kilo-* for a thousand. I've already used "kilogauss" for a thousand gauss, assuming that the prefix is familiar to you from such terms as *kilometer* and *kilogram*. Similarly, *milli-* is commonly used for thousandths, as in *millimeter* and *milligram*. With the exception of the familiar *centi-* (for one hundredth) and the less common *deci-* (for one tenth) and *deca-* (for ten), the prefixes refer to factors of a thousand. For example, although *micro-* and *mega-*, borrowed from Greek, are generally used for "small" and "big" (*microscope, megaphone*), in science they are used specifically for "millionths" and "millions."

Once you become familiar with these prefixes, they have a way of creeping into nonscientific usage. Annual salaries are often expressed in terms of "kilobucks," university budgets in "megabucks," and government programs in "gigabucks." Unfortunately, our national debt now can be measured in *terabucks!*

NUMBER NAMES

Quantity	The long way	Scientific notation	Prefix
Quadrillionth	0.000000000000001	10^{-15}	femto-
Trillionth	0.000000000001	10^{-12}	pico-
Billionth	0.000000001	10^{-9}	nano-
Millionth	0.000001	10^{-6}	micro-
Thousandth	0.001	10^{-3}	milli-
Thousand	1,000	10^{3}	kilo-
Million	1,000,000	10^{6}	mega-
Billion	1,000,000,000	10^{9}	giga-
Trillion	1,000,000,000,000	10^{12}	tera-
Quadrillion	1,000,000,000,000,000	10^{15}	peta-

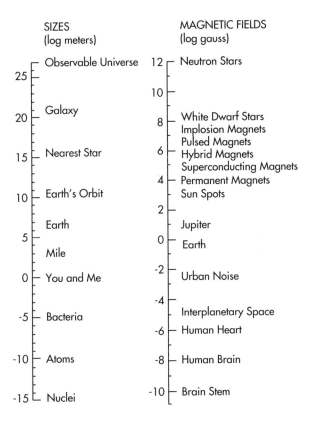

Figure 4.3 A comparison of the size *(left)* and magnetic field *(right)* of different objects. The numbers in a log scale like those used here refer to the power of ten in the units specified (meters or gauss, as indicated); for example, an atomic nucleus is 10^{-15} meters in size, and the magnetic field of a neutron star is 10^{12} gauss.

To compare graphically a series of very small and very large quantities, we use a "logarithmic" scale, in which the exponent or logarithm, with 10 as the number base, is displayed linearly along an axis. The logarithm of 1,000 is 3, and the logarithm of 1,000,000 is 6. The most familiar logarithmic scale, at least to Californians, is the Richter scale, used to measure the magnitude of earthquakes. Each increase of 1 on the Richter scale corresponds to a tenfold increase (an addition of one zero) in intensity. A quake of magnitude 6 is therefore a *thousand* times more intense than a quake of magnitude 3. Another example of logarithmic increase is a gradation of the sizes of things (Figure 4.3), from the diameter of atomic nuclei to the size of the observable universe. Each mark along the axis represents another factor of 10 in size. Similarly, the strength of various magnetic fields that we have discussed so far, or will encounter in later chapters, is illustrated on a logarithmic scale.

As the figure shows, the magnetic fields in neutron stars, discussed in Chapter 3, are a trillion times stronger than the earth's magnetic field, and some of the magnetic fields generated by brain activity, to be discussed in Chapter 15, are less than a billionth of the earth's field. And yet, via twentieth-century advances in physics, we can now measure both.

///// SUPERCONDUCTING MAGNETS

5

Amps, Volts, and Ohms

The units we use to measure electricity and the electrical properties of materials commemorate three European physicists who pioneered research in this area. Electrical current is measured in amperes, after André-Marie Ampère (1775–1836), the French physicist who, shortly after hearing the news of Oersted's 1820 discovery that an electric current moved a compass needle, developed a mathematical theory of electromagnetism. Voltage is measured in volts, after Alessandro Volta (1745–1827), the Italian physicist whose invention of the electric battery in 1800 provided the first source of continuous electric current. And electrical resistance is measured in ohms, after Georg Simon Ohm (1789–1854), the German physicist who discovered that the current flow through a conductor (a foot of copper wire, say) was proportional to the voltage difference across that conductor (double the voltage difference and you'll double the current). This is Ohm's law, and the voltage difference across the wire (in volts) divided by the current (in amps) is called the electrical *resistance* (in ohms).

I perhaps should remind you of the difference between current and voltage. The flow of electricity through wires is often likened to the flow of water through pipes, with the electric *current* (amps) analogous to the rate of water flow (gallons per minute, say). In this picture, *voltage* is analogous to what is pushing the water, the water pressure (in pounds per square inch, say). Ohm's law applied to water flow would say that the rate of water flow was proportional to the water pressure. A more pleasant image is that of water flowing down a mountain stream. Current is still represented by the rate of water flow, but now the "voltage" driving the current is the difference in altitude between the mountaintop and the sea toward which the water is flowing. Here gravity provides the pressure.

You might object to this analogy by saying that flowing water moves whereas a current-carrying wire doesn't. To explain what is "moving" in an electric current, we must examine the wire at the atomic level. The atomic number of copper, a common conducting material, is 29. An isolated copper atom consists of a dense nucleus containing 29 positively charged protons (and some neutrons) surrounded by a diffuse cloud of 29 negatively charged electrons. When this copper atom bonds with other copper atoms to form a solid like our copper wire, however, there is a small but very important change in atomic structure. Of those 29 electrons per atom, 28 are homebodies—they remain within the confines of their own atoms, held by electrical attraction to the positive charge of the nucleus. But one electron per atom becomes a wanderer, moving freely through the wire, shared by all the atoms. These are called "free" electrons, and they wander randomly in all directions unless a voltage difference is applied across the wire. An applied voltage, however, produces a force on the electrons that leads to a net flow of the free electrons in one direction. This net flow of free electrons is electric current. One ampere of current corresponds to almost 10^{19} (10 billion billion) electrons per second flowing across each cross section of the wire. (If you've applied the voltage by connecting the two ends of your wire to the two terminals of a battery, the electrons will enter the wire from the negative terminal and exit the wire into the positive terminal. Joining just one end of the wire to the battery won't give you current, since the electrons need a complete loop to circumnavigate.)

Metals typically have between one and three free electrons per atom, and so they conduct electricity quite well. The nature of interatomic bonding is very different in other materials. Some materials (insulators, like most glasses, plastics, and ceramics) have essentially no free electrons, and some materials (semiconductors) have only a few. In transistors and integrated circuits, a semiconductor can do a great deal with, for example, only one free electron per 10,000 atoms. But if you want to carry large amounts of electrical current, you'd better use a metal, and copper's a good choice.

Water molecules flowing through a pipe bump into each other, and into the walls of the pipe, creating friction. Without a steady water pressure pushing them along, the water molecules would stop flowing. Free electrons flowing through a wire bump into the nuclei and all the other electrons (most of which, you'll remember, stay at their home atoms), creating friction. Without a steady voltage difference pushing them along, the current would stop flowing. This frictional resistance to the flow of free electrons is our atomic-level view of electrical resistance. Because of this friction, current flow heats up the wire and requires the delivery of energy from the battery (or other source of voltage) to make up the energy lost in heating. Electrical resistance and the associated heating is not always a bad thing. We make use of it in space heaters, toasters, and incandescent lamps. But it was the heating produced by electrical resistance that required Francis Bitter to use so much electrical power and so much cooling water in his high-field electromagnets. It would have been so much easier to produce high-field electromagnets if there were a material in which electrons could flow without friction—without heating—without energy consumption—without electrical resistance—ohmless.

In scientific research, as in earth or space exploration, you maximize your chances of discovering something new if you "boldly go where no one has gone before." Columbus traversed the Atlantic, Captains Kirk and Picard comb the galaxies, and Dutch physicist Heike Kamerlingh Onnes measured the properties of materials at temperatures colder than anyone had reached before.

In the early years of this century, Onnes and other physicists were exploring lower and lower temperatures. As you go to lower and lower temperatures, gases turn to liquids, liquids turn to solids, and, within solids, the random vibrational motion of atoms becomes less and less (when you chill, the atoms still). In 1908, Onnes and his research group at the University of Leiden became the first to reach a temperature so low that even helium—the most stubborn of the gases—became liquid.

With the production of liquid helium, the laboratory at Leiden became "the coldest spot on earth." (In the supercold or "cryogenic" regime, temperatures are usually measured in degrees Kelvin, with zero set at "absolute zero," the lowest temperature even theoretically attainable. The temperature of liquid helium was only about 4 degrees Kelvin, 269 degrees below zero Centigrade or 450 degrees below zero Fahrenheit.) Once Onnes could explore the properties of materials at record-breaking temperatures, he decided to measure, among other things, the electrical properties of metals. He passed electric current through various metals and measured changes in the electrical resistance (the ratio of voltage to current) as temperature decreased.

As temperature decreased, Onnes found that the resistance of various metals gradually decreased. This was no surprise, because it was already known that the vibrational motion of atoms contributed to electrical resistance, and this motion decreased as temperature decreased. The surprise came in 1911, when he cooled a wire of mercury below the temperature of liquid helium. The electrical resistance dropped abruptly to *zero!* Current continued to flow through the wire, but there was no voltage. Onnes had discovered a remarkable phenomenon that he named *superconductivity.*

///// Superconductors and Ohmless Electromagnets

In further experiments, the group at Leiden learned that mercury was not the only metal that lost its electrical resistance—became superconducting. In tin, lead, indium, and other metals, resistance also dropped to zero abruptly at a *critical temperature.* The critical temperatures were different for each metal, but all were very low—only a few degrees above absolute zero.

Onnes had learned that, when cooled below their critical temperatures, several metals could conduct electric current with no voltage—no resistive heating—no loss of energy. Something had happened below the critical temperature that allowed electrons to flow without friction. What can you do with such materials? He realized that the electrical resistance of copper, and the associated heating and energy loss, limited the magnetic fields attainable with ordinary electromagnets. Since superconductors have no electrical resistance, why not build an electromagnet with superconducting wire? In this way you should be able to use huge currents to create huge magnetic fields. Onnes wrote in 1913:

> The extraordinary character of this state [superconductivity] can be well elucidated by its bearing on the problem of producing intense magnetic fields with the aid of coils without iron cores. Theoretically it will be possible to obtain a field as intense as we wish by arranging a sufficient number of ampere windings round

the space where the field has to be established [a sufficient number of turns in the coil of an electromagnet] . . . a field of 100,000 gauss could then be obtained by a coil of say 30 centimeters in diameter and the cooling with helium would require a plant which could be realized in Leiden with a relatively modest financial support.

Onnes received the financial support he was fishing for but never attained his goal of a 100,000-gauss electromagnet. Unlike many scientists looking for financial support, Onnes had openly foreseen the possibility of failure: "There remains of course the possibility that a resistance is developed in the superconductor by the magnetic field." Unfortunately for him, that's how it turned out. The metals he studied all lost their superconductivity at magnetic fields of only a few hundred gauss.

Onnes learned that superconductivity disappeared not only when the temperature exceeded the critical temperature but also if the magnetic field exceeded a *critical field,* and if the current exceeded a *critical current.* The phenomenon of superconductivity was remarkable, but it was also delicate. It disappeared if the temperature, magnetic field, or electric current were too high. To maintain superconductivity, those three quantities—temperature, field, and current—must all be kept below critical values characteristic for each material. To produce 100 kilogauss in a superconducting magnet, the material must first be cooled well below its critical temperature. In addition, at the temperature of operation, the material must not only have a critical field above 100 kilogauss, it must also have a high critical current in the presence of 100 kilogauss. Before these challenges were met, half a century passed. It was not until 1961 that scientists at the Bell Laboratories reported that a compound of niobium and tin was capable of carrying large superconducting currents even in the presence of a large magnetic field. Soon thereafter, Onnes's goal of a 100 kilogauss superconducting electromagnet was achieved at General Electric—another materials breakthrough I was fortunate to witness firsthand.

We now know that many alloys and compounds have very high critical fields and can have high critical currents, but Onnes and the physicists who followed him confined their studies for many years mostly to pure metals, which do not. It was the scientists at Bell, GE, and other industrial laboratories who pioneered the development of high-field, high-current superconductors and soon produced commercial high-field superconducting magnets. This was a development that required the close cooperation of physicists, materials scientists, and engineers. It was an exciting time.

In Schenectady with General Electric, I worked for several years on the materials science of high-field superconductors. I and my colleagues around the world learned, through experiments, the relations between the manufacture, microstructure, and properties of these promising new materials. We found that the critical current, a property crucial to the construction of high-field magnets, was extremely dependent on the superconductor's microstructure—on how the atoms of the alloy or compound were arranged at the microscopic level. Two wires of the same alloy, looking exactly the same to the naked eye, could differ by factors of many thousands in their critical currents if, because of differences in processing, they had different microstructures. The relation between microstructure and critical current of superconductors is closely analogous to the relation between microstructure and coercivity of permanent magnets, both topics of the next chapter.

Today most superconducting magnets produce fields between 50 and 90 kilogauss and are wound with niobium-titanium, a ductile alloy that can be easily drawn into the miles-long wires that are required. Superconducting magnets of higher fields, up to about 200 kilogauss, are wound with niobium-tin, a brittle compound that requires more complex processing techniques and is considerably more expensive than niobium-titanium.

Although the importance of the Bitter magnets at the National

Magnet Lab was diminished by the development of superconducting magnets, they were very useful to the researchers developing high-field superconducting materials, since they allowed measurement of critical currents of their conductors in high magnetic fields before the superconducting magnets became readily available. Bitter magnets have also been combined with superconducting magnets to produce large "hybrid" magnets that have generated record magnetic fields. In 1994, a Bitter magnet surrounded by niobium-tin and niobium-titanium superconducting magnets, and enhanced by a central core of ferromagnetic holmium, a high-magnetization rare-earth metal, generated a steady magnetic field of 387 kilogauss. (Holmium becomes ferromagnetic at low temperatures and has a higher saturation magnetization than iron.) The outer superconducting electromagnets produced 122 kilogauss, the inner Bitter magnet added 230 kilogauss, and the holmium contributed 35 kilogauss. This record surpasses the lab's 1992 record of 372 kilogauss, which had been recognized by the *Guinness Book of Records* as the world's highest continuous field. (Pulsed magnets, you may recall, can produce much higher fields, but only for very short times. Continuous fields are much more useful to experimenters.)

Workers from MIT's magnet lab are now collaborating with the new National High Magnetic Field Laboratory in Florida in the design of a more powerful hybrid magnet to be constructed in Tallahassee. The inner Bitter-style magnet will generate 310 kilogauss, using 24 megawatts of electrical power. That magnet will be surrounded by superconducting magnets producing an additional 140 kilogauss, for a total steady central magnetic field of 450 kilogauss, a million times the strength of the earth's field! As with the earlier hybrids, the outer superconducting electromagnet is wound with niobium-titanium. Inside that, where the fields are higher, the superconductor used is niobium-tin, which has a higher critical field than niobium-titanium. And inside the niobium-tin magnets is the Bitter magnet, which is limited not by critical fields but by power and cooling needs and the tremendous forces created by such huge fields. (A holmium insert will probably be added for the sake of the record books, but will not be present in normal operation because it consumes space required for experimental apparatus. These magnets are designed for research into the properties of materials in high magnetic fields, not for Guinness.)

///// Big and Little Science

Prior to 1961, superconductivity was little more than a fascinating scientific phenomenon, with no important applications. After the development of high-field superconducting wires using niobium-titanium and niobium-tin, it became an important techno-

MAGNETS FOR THE RECORD

The *Guinness Book of Records* was first published in 1955, and by 1989 updated editions had sold 60 million copies, second only to the Bible on the all-time bestseller list. MIT's National Magnet Lab has two entries: the world's highest steady field, produced by the hybrid magnet, and the world's lowest magnetic field, produced in a magnetically shielded room used for measurements of magnetic fields generated by the human brain. Other Guinness records related to magnets include:

- the world's largest electromagnet, at CERN, a European lab for nuclear research located near Geneva, Switzerland;

- the world's most powerful particle accelerator, the Tevatron at Fermilab, near Batavia, Illinois, which uses over 1,000 large superconducting magnets; and

- the world's lowest temperature, only a tiny fraction of a degree above absolute zero, achieved in Helsinki, Finland, using high magnetic fields. (Magnets can also be used to super-refrigerate!)

logical tool, in the form of high-field superconducting electro-magnets.

The major commercial use of superconducting magnets today is in magnetic resonance imaging (MRI), which is discussed in Chapter 14. Other commercial applications have included energy storage (storing electric energy that can be drawn upon in the case of short-term power outages) and magnetic separation for the purification of clays. Demonstration projects using supercon-ducting coils have included electric generators, magnetic levita-tion of high-speed trains (Chapter 10), and ship propulsion (an experimental ship propelled by superconducting "thrusters" was launched in Japan in 1990). But these various applications have so far been overshadowed by the use of superconducting mag-nets in basic scientific research.

Small superconducting magnets have become a standard item of laboratory equipment. Individual scientists throughout the world use them to study the effects of high magnetic fields on various physical, chemical, and biological phenomena. Although small in comparison with the giant magnets used in particle ac-celerators and fusion research, laboratory superconducting mag-nets are typically contained in metal "thermos bottles" several feet high and perhaps two feet in diameter. But they produce magnetic fields of far greater strength, and far greater extent, than most people can imagine, as the following story attests. Although warning signs are usually posted when laboratory magnets are in operation, English is often not the primary lan-guage of workers in night cleaning crews. One such worker entered the laboratory of an acquaintance of mine and, unaware of the presence of large but invisible magnetic fields, got a bit too close to an operating superconducting magnet. Several items of cleaning equipment, containing steel parts, suddenly flew through the air and attached themselves to the metal cylinder housing the magnet. The worker ran screaming from the room, convinced that poltergeists had taken over the laboratory.

Superconducting magnets have also had a tremendous impact

on several large-scale research projects involving hundreds or even thousands of scientists, projects that have come to be called "Big Science" (the work of individuals or small groups of researchers is analogously called "little science"). Two Big Science areas affected most strongly by superconducting magnets are high-energy physics and fusion. In each, high magnetic fields are needed to control the motion of charged particles. This discussion requires the introduction of another fact about magnetic forces, an important and fascinating fact but one that involves some three-dimensional thinking. (In freshman physics courses, fact 9 is presented in the form of an equation and is used as the definition of a magnetic field. Electric fields are defined in terms of the force they exert on stationary charged particles, and magnetic fields are defined in terms of the force they exert on *moving* charged particles.)

> *Fact 9:* A charged particle experiences no magnetic force when moving parallel to a magnetic field, but when it is moving perpendicular to the field it experiences a force perpendicular to both the field and the direction of motion.

The first part is easy. A charged particle moving parallel to a magnetic field (Figure 5.1A) feels no force and therefore continues to move in a straight line. Easy, but rather dull.

The second part, though harder to visualize, is far more interesting. A particle moving in a direction perpendicular to the field (Figure 5.1B) feels a force perpendicular to its direction of motion. This deflects its path, but the force always remains perpendicular to the motion, and the particle moves *in a circle.* (An analog is the circular motion of the moon around the earth, where the attractive gravitational force remains perpendicular to the moon's direction of motion.) The radius of the circle decreases with increasing field, but increases with increasing speed of the particle.

What if the particle's motion is neither exactly parallel to, nor

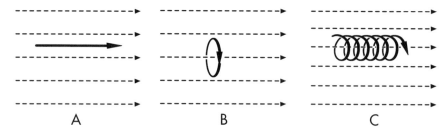

Figure 5.1 The trajectories of electrons moving at various angles to a magnetic field. (A) An electron moving parallel to the field experiences no force and continues moving in a straight line. (B) An electron moving perpendicular to the field experiences a perpendicular force that produces a circular motion in a plane perpendicular to the field. (C) An electron moving at an intermediate angle to the field will spiral along the field.

exactly perpendicular to, the field (Figure 5.1C)? You can look at this general case as the sum of the two previous cases. The motion along the field remains uniform and unchanged, as in Figure 5.1A, but the motion across the field remains circular, as in Figure 5.1B. Adding the two together, the net result will be helical, or screw-like, motion along the field (Figure 5.1C).

In a previous chapter, I mentioned the aurora borealis, the display of light in earth's atmosphere produced by charged particles emitted by the sun. Once the particles enter the earth's magnetic field, they travel helically along the field lines, as in Figure 5.1C. Since the earth's field lines run from pole to pole, the particles will enter the atmosphere, and excite atoms of the atmospheric gases into emitting light, when the charged particles reach the polar regions (Figure 5.2). That's why the northern lights are so far north!

High-energy physicists enjoy accelerating electrons, protons, and other charged particles up to very high speeds and then crashing them into various atoms, or into each other, to see what happens. In this way, they have learned much about the internal structure of atomic nuclei and subatomic particles and about the mysterious forces that hold matter together. The highest energies, and many of the most exciting discoveries, have been made in machines that use high magnetic fields to keep charged particles traveling over circular paths (Figure 5.1B) for many laps as they are accelerated (by electric fields) to higher and higher energies. The current record holder, as noted above, is located at the Fermilab in Illinois. It's called the Tevatron. (The Tevatron gets its name from TeV, an abbreviation for tera-electron-volts—one trillion electron volts, the beam energy for which it was designed. An electron volt is a unit of energy—the energy gained by an

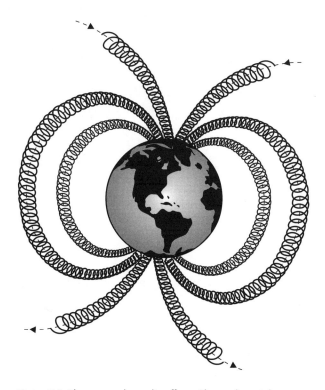

Figure 5.2 The aurora borealis effect. Charged particles are emitted from the sun and travel through space. When they reach the earth's field, their path is changed by the field (as in Figure 5.1C). Particles spiral along the magnetic field lines and enter the earth's atmosphere near the poles, creating auroral displays.

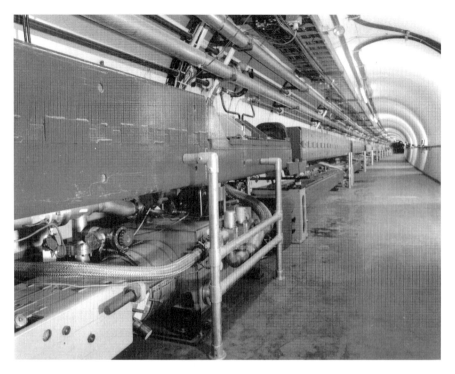

Figure 5.3 The four-mile main accelerator at Fermilab (Batavia, Illinois) is located in a tunnel 20 feet below the ground. The upper ring consists of copper-wound electromagnets. The lower ring is the Tevatron, composed of superconducting magnets designed to accelerate protons up to energies approaching 1 TeV (one trillion electron volts).

electron or proton when accelerated through a potential difference of one volt.)

The original accelerator at Fermilab used "conventional" (not superconducting) electromagnets to guide and focus proton beams. The Tevatron, installed in the same 4-mile-long circular tunnel (Figure 5.3), contains over one thousand superconducting magnets, averaging about 20 feet long. Most are "dipole" magnets, creating vertical magnetic fields that peak at 45 kilogauss and bend the particle trajectories into a circular path. Others, called "quadrupoles," create a more complex arrangement of magnetic fields produced by four magnetic poles and are used to focus the particle beams. The superconducting magnets have enabled the Tevatron to accelerate proton beams up to energies of 900 billion electron volts, more than twice the energies achieved in the ring with conventional magnets, yet with significantly lower operating costs. Kamerlingh Onnes would be proud!

The thousands of superconducting electromagnets in the Tevatron ring and the conventional electromagnets in the "Main Ring" (the upper ring in Figure 5.3) are by no means the only magnets in the Fermilab facility. A "small" accelerator called the Booster, 500 feet in diameter, uses magnets to keep charged particles in a circular path while electric fields boost their energy up to 8 billion electron volts before they are directed into the Main Ring. And many magnets are used in the huge (many thousands of tons) detectors that analyze what happens when the high-energy particle beams collide with either fixed targets or oppositely directed particle beams. The data gathered by tons and tons of giant magnets are then stored in infinitesimal magnetic particles in magnetic recording tape for later analysis. In April 1994, Tevatron researchers announced the tentative identification of the "top quark," the elusive, short-lived particle that had been a key missing piece of the current theory of the structure of matter. Nobody needs magnets, and fact 9, more than researchers in high-energy particle physics.

Several years of successful results with the Tevatron led directly to plans for the now-infamous SSC—the Superconducting SuperCollider. This, while it lasted, was really BIG Science! The SSC was to use over ten thousand superconducting magnets, averaging over 50 feet long but with some over 90 feet long, in a ring over 50 miles long. The magnets, wound with niobium-titanium, were to supply a field of 66 kilogauss, and two counter-rotating proton beams were to be accelerated to an energy of 20 TeV and then brought into head-on collision with each other.

Construction on the SSC began in 1989. The site chosen was

tiny Waxahatchie, Texas. President George Bush, an enthusiastic supporter and part-time Texan, said in 1992, "This is the Louvre, the Pyramids, Niagara Falls all rolled into one." (Even for an American politician, that was a bit of an exaggeration. But those earlier "construction projects" were completed.) That August, the current in a string of five completed magnets was successfully increased to 6,250 amps, reaching the intended magnetic field of 66 kilogauss. But Congress wasn't sufficiently impressed. Steadily increasing cost estimates and a growing concern about the federal budget deficit finally killed the SSC in the fall of 1993, after about 2 billion dollars had already been spent, and after 11 miles of tunnel had already been dug under Waxahatchie. It was a Big Setback for Big Science, but the failure was political and financial, not technical. The superconducting magnets had shown they could handle the job.

The hopes of the high-energy physics community are now centered at CERN, the laboratory of the European Organization for Nuclear Research that spans the French-Swiss border near Geneva. In a 16-mile circular tunnel that already houses LEP, the Large Electron-Positron Collider, CERN plans to build LHC, the Large Hadron Collider. (Positrons, positively charged particles with the same mass as electrons, will reappear in later chapters. Hadrons are any of a group of heavy particles, which includes the proton and a lot of even heavier particles with Greek names.) Although the LEP uses non-superconducting magnets, the LHC will use niobium-titanium superconducting magnets, the dipole magnets designed to reach 100 kilogauss. For colliding proton beams, the collision energy could reach 14 TeV, and for collisions of heavier ions, much higher collision energies should be attained. It is an ambitious program, but if the countries of Europe do not turn off the cash flow, the LHC should be operating by 2003, and the next discoveries in high-energy physics will be coming from Europe instead of Texas.

Another Big Science area that has had trouble maintaining

government funding is fusion. As noted in an earlier chapter, fusion, the merging of light atoms into heavier ones, is the mechanism used by the sun to generate energy. Controlled thermonuclear fusion has been touted for many years as a potential source of cheap, clean energy. It involves the containment of a plasma (an ionized gas) at temperatures of 100 million degrees. (This is clearly not *cold* fusion, the now-discredited claim by two electrochemists that they had achieved fusion in a small chemical cell.) The ions in the plasma, being charged, experience forces from magnetic fields via fact 9. Several devices employing large electromagnets to "contain" the plasma—to keep the high-energy, fast-moving ions together long enough to achieve fusion—have been built and studied.

A "tandem-mirror" facility at Livermore, California, was constructed and tested in 1986. It required 42 superconducting magnets, including two huge end magnets that weighed 340 tons each. Using both niobium-titanium and niobium-tin conductors, the project successfully reached the design fields, which were as high as 125 kilogauss. Shortly after final assembly and test, however, funding was cut off and the project mothballed. If the people at Livermore need somewhere to store their now-unused magnet assembly, I know some people in Texas who have a spare tunnel.

A design now considered more practical for a fusion reactor is a doughnut-shaped magnet called a tokamak. A large tokamak at Princeton, using conventional magnets, in June 1994 set a record for terrestrial power generation by fusion—9 million watts (enough to power a town of about 15,000 people). But this power level was only maintained for less than half a second, and the machine still consumed more than twice as much power as it generated. Future tokamaks will use superconducting magnets, and to assist in developing the necessary technology, an international program was organized to install six large (40 ton) tokamak-style superconducting coils at the Large Coil Test Facility at Oak Ridge, Tennessee, in 1986. Coils were produced in the United States, Europe, and Japan, using different materials and different conductor geometries, and each exceeded its design field of 80 kilogauss.

Superconducting tokamaks have been built in Europe and Japan. But the most ambitious fusion project on earth is now the International Thermonuclear Experimental Reactor (ITER) being designed by a team that includes the European Atomic Energy Community, the United States, Japan, and Russia. Design activity is underway at sites in San Diego, California, in Garching, Germany, and in Naka, Japan. (Time differences are such that someone is working on ITER around the clock!) Since design fields reach 130 kilogauss, which is beyond the capability of niobium-titanium, niobium-tin must be used for much of the magnet construction. Whereas the Princeton tokamak sustained fusion for less than a second, the ITER device has the objective of an extended "burn" of at least 15 minutes. If successful, it could be the last test reactor before the construction of a device capable of practical generation of power by fusion. Most experts estimate that practical fusion power remains at least 25 years in the future. ("Fusion is the energy source of tomorrow, and it always will be," is a standard joke in Washington.) But when it arrives, it will depend on superconducting magnets and fact 9.

///// The Woodstock of Physics

Although superconductivity was discovered by Onnes in 1911, it was not "discovered" by *The New York Times* or most other media until 1987. Until the previous year, all known superconductors had critical temperatures less than about 20 Kelvin—20 degrees above absolute zero. The coils of superconducting magnets were cooled by liquid helium, which is expensive and difficult to handle. In 1986, two scientists at the IBM lab in Zurich, J. Georg Bednorz and K. Alex Müller, found a complex oxide material with a critical temperature of about 30 Kelvin. Their discovery triggered intense scientific activity worldwide, which soon led to the discovery of superconducting oxides with critical temperatures over 90 Kelvin. These new oxides were superconducting when immersed in liquid nitrogen (77 Kelvin), which is much cheaper and easier to handle than liquid helium. Since superconductivity was much easier to achieve in these materials, major technological advances seemed possible.

The newly discovered "high-temperature" superconductors generated both tremendous interest among scientists and a frenzy in the popular press. Magazine cover stories and front-page articles in major newspapers covered each step and misstep in attempts at further progress. Headlines screamed about "momentous advances," "profound developments," "major breakthroughs," and a potential "multibillion dollar business." Physicist Paul Chu of the University of Houston was quoted in *Time* magazine as saying: "It could almost be like the discovery of electricity"—a typical Texas overstatement, but it represented the feeling of many scientists in 1987. President Reagan, citing the threat of foreign competition, announced a national superconductivity plan. Premature claims of superconductivity at much higher temperatures, including room temperature and above, appeared almost weekly, only to be withdrawn soon afterward. Bednorz and Müller were awarded the 1987 Nobel Prize in Physics, an unusually rapid recognition of a scientific accomplishment.

At the March 1987 meeting of the American Physical Society in New York, a session on high-temperature superconductors drew thousands of enthusiastic and curious scientists, myself included. The huge ballroom in the Hilton was not huge enough for this crowd, so TV screens were set up in other rooms to accommodate the overflow. The session opened a bit after 7 P.M., and went on and on and on, as speaker after speaker reported their latest results. One speaker announced "Our lives have changed." After midnight and five hours of talks had passed, the crowd finally began to thin. By about 3 A.M., I wore out and returned to my hotel room. I was told that a small knot of enthusiasts was still there at dawn. The newspapers dubbed the event "the Woodstock of physics."

As I write this in 1995, high-temperature superconductivity has long been off the front pages, but scientific activity remains strong worldwide. Maximum critical temperatures have exceeded 130 Kelvin, but in Fahrenheit, this is still more than 200 degrees below zero, not a "high temperature" in laymen's terms. There is no question that the Bednorz-Müller discovery triggered a scientific breakthrough, but whether it will lead to any major technological breakthrough remains to be seen. The new materials have high critical temperatures and high critical fields, but critical currents at high fields remain disappointing. They are not yet sufficient to construct high-field electromagnets.

The intensity of research effort in university, industrial, and government laboratories across the world will undoubtedly lead to further improvements in critical currents and to some commercial applications of high-temperature superconducting oxides. But from today's perspective, the major breakthrough in the development of superconducting magnets remains the 1961 Bell Labs discovery of high-field, high-current superconductivity in niobium-tin.

I explained earlier that the critical current of high-field superconductors, crucially important for the construction of superconducting electromagnets, is very sensitive to the microstructure within the superconductor. The technogically important properties of permanent-magnet materials, and of the temporary magnet materials used in ordinary (non-superconducting) electromagnets, are also very sensitive to microstructure. To get a glimpse of why the engineering properties of magnetic and superconducting materials are so sensitive to microstructure, we have to take a look *inside* the materials. We do that in the next chapter.

6

///// In a Spin

Physicists will tell you that, strictly speaking, all matter is magnetic. With sufficiently sensitive measuring equipment, you can measure the magnetic properties of copper, salt, glass, plastic, and even living things—you're magnetic. But most animate and inanimate matter is only very weakly magnetic. Without sensitive equipment, we could never detect its magnetism.

Some matter is very different. Einstein's compass needle was magnetic enough to swing around in response to the earth's magnetic field. The alnico magnets under my Scottie dogs were magnetic enough to exert forces on each other, or on paper clips and refrigerators, that were easy to see. No sensitive equipment was needed.

Why are iron and steel and a few other things strongly magnetic when most matter is not? Why could Gilbert and Sullivan's magnet attract knives and needles but not a silver churn? Why are lodestones permanent magnets, while pure iron is a temporary magnet?

To begin to answer such questions, we have to start at the scale of atoms ("think small"), because most of the magnetism of matter is generated by *electrons*. Although atomic nuclei are weakly magnetic, a fact we make use of in MRI (Chapter 14), the magnetism of matter is mostly electron magnetism. When electric current flows through the copper wire of an ordinary electromagnet, or through the niobium-titanium wire of a superconducting magnet, it creates a magnetic field (fact 6). Since electric current in metals is simply the motion of electrons through the metal, we know that the motion of electrons produces a magnetic field. We've known that since 1820, thanks to Oersted. But electric current is long-distance motion of electrons, and, as noted in Chapter 1, permanent and temporary magnets produce magnetic fields even when no electric current is flowing. In some materials, electrons can produce magnetism strong enough to hold notes to your refrigerator without ever leaving their home atoms!

Simplified models of the atom describe it as a tiny solar system, with the electrons revolving about the nucleus like planets around the sun and, again like the planets, also rotating about their own axes. Both electron motions produce magnetism, but the first component, called "orbital magnetism," is not important in most solids. In iron and most strongly magnetic materials, it is the second component of electron motion, the electron's *spin* about its own axis, that creates most of the magnetism.

If an atom with one spinning electron is magnetic, then an atom with two spinning electrons should be more magnetic, right? Wrong! An atom of helium (He) and a diatomic molecule of hydrogen (H_2) each has two electrons. But an interaction force between the spins of neighboring electrons called *exchange* (related to the indistinguishability of electrons, so that nothing changes if two electrons exchange places) makes the two electrons in the helium atom and the hydrogen molecule spin in opposite directions. Net spin magnetism, therefore, is zero.

This exchange force between electron spins cancels the electron spin magnetism of most electrons of most atoms. An atom of radon contains 86 electrons, but 43 electron spins point one way, and 43 point the opposite way. Net magnetism, zero. Luckily for modern technology, and therefore for us, not all atoms are so unbiased.

Elements like radon, in which all the electron spins cancel (or nearly cancel), are weakly repelled by a permanent magnet and are called *diamagnetic*. (Silver is diamagnetic, so Gilbert and Sullivan's silver churn was actually slightly repelled, rather than attracted, by the magnet.) In many other materials, the electron spins on each atom do not completely cancel, and so each atom may be considered a very tiny magnet. Yet most of these materials remain only very weakly magnetic, because the individual atomic magnets remain independent of their neighbors and point in random directions. Such materials, only slightly attracted to a permanent magnet, are called *paramagnetic*.

In a few materials, however, the net spins of neighboring atoms are coupled strongly by the exchange force. This "exchange coupling" between neighboring atomic magnets can be either positive or negative, producing parallel or antiparallel spins. In chromium and manganese, each atom is strongly magnetic, but neighboring atomic magnets are forced by exchange to point in *opposite* directions. Such materials are called *antiferromagnetic,* and they have little net magnetization. The magnetic properties of most materials—diamagnetic, paramagnetic, or antiferromagnetic—are so weak that, in common parlance, they are considered nonmagnetic. They do not have "the right stuff" to be useful as permanent or temporary magnets.

In *ferromagnetic* iron, cobalt, and nickel, however, the exchange force between neighboring atomic magnets makes them point in the *same* direction. Rather than cancelling each other out, all the atomic magnets therefore can add together to produce a substantial macroscopic magnetization. The wonderful macroscopic effects that we observe with permanent and temporary magnets result from many, many individual atomic magnets acting in unison. Individual iron atoms are magnetic because of unbalanced electron spins, and the individual strengths of each are summed together via the positive exchange force that produces parallel coupling of net spins on neighboring atoms. Ferromagnetism (from the Latin word for iron) is a phenomenon of large-scale interatomic cooperation.

Of over 100 elements in the periodic table, only these three—iron, cobalt, and nickel—are ferromagnetic at room temperature. (Gadolinium has a Curie temperature slightly below room temperature, and other rare-earth metals, like holmium, become ferromagnetic at still lower temperatures.) Compared with iron, cobalt and nickel are relatively rare and relatively expensive. We are fortunate indeed that iron, the most common metal in the universe and in our planet, and a metal easily and inexpensively produced from common ores, is ferromagnetic. It is difficult to imagine how different our technological world would be if the Curie temperature of iron were below room temperature.

///// The Magnetic Domain

Permanent magnets are popular subjects in elementary-school science. In the upper grades, or perhaps in high school, students may do or see a few experiments with electromagnets. But college physics courses rarely try to explain ferromagnetism; perhaps you can already see why. To answer the simple question of why iron is ferromagnetic, I've had to use the strange concept of electron spin and an exchange force (of quantum-mechanical origin) that couples spins—sometimes in opposite directions, sometimes in the same directions. (Simple questions don't always have simple answers. A small child can ask why grass is green, but it takes me several lectures to explain that to my students at MIT.)

Properly presented, spin and exchange are topics for advanced courses in quantum mechanics. For our purposes, it's enough to know that each iron atom is a tiny magnet because the spins of its 26 electrons do not completely cancel and that iron is ferromagnetic because the net spins of neighboring iron atoms point in the same direction. But I still have a bit of explaining to do. In particular, I need one more important idea to explain the difference between a refrigerator magnet and a refrigerator. One is a permanent magnet, the other is a temporary magnet. Why?

It's true that a permanent magnet, or a temporary magnet that's been magnetized by a permanent magnet or an electromagnet, has all (or nearly all) its atomic magnets pointing in the same direction. That gives it a net macroscopic magnetization, north and south poles at its ends, and a magnetic field extending from it that you can use to pick up paper clips or perhaps a villain with steel teeth. In the magnet illustrated in Figure 6.1A, an arrow represents the total magnetic strength of all the atomic magnets added together. (That's the quantity we called the *saturation magnetization* back in Chapter 2.) But temporary magnets are only temporary. When we turn off the current to the electromagnet, or remove the permanent magnet, how can the temporary magnet lose its magnetism? The electrons are still spinning, the exchange force hasn't been turned off. A French physicist named Pierre Weiss answered this question in 1929, by introducing the concept of *magnetic domains.*

A demagnetized iron bar, like the one in Figure 6.1C, may be divided into four magnetic domains, for example. Two domains are magnetized up, two down. Net macroscopic magnetization, zero. Instead of a north pole at one end and a south pole at the other, and a far-reaching external magnetic field (as in Figure 1.1A), each end has a mix of north and south poles, and external magnetic fields are very limited in extent. But the atoms are still magnets, and most of them are still exchange-coupled to point in the same direction as their neighboring atoms. Within each domain, the iron is still fully magnetized to its saturation magnetization, but each domain is magnetized in a different direction, represented by the arrows. The only "unhappy" atoms (atoms with spins not parallel to those of their neighbors, the exchange force therefore not fully satisfied) are the few along the boundaries between the domains, across which the direction of magnetization changes. These boundaries are called *domain walls*, and controlling the behavior of magnetic domain walls is the major task of materials scientists who work with magnetic materials, whether permanent-magnet or temporary-magnet materials.

A word about terminology. Engineers and scientists usually refer to permanent-magnet materials and temporary-magnet materials as "hard" and "soft" magnetic materials, respectively. This terminology arose over a century ago from observations that mechanical hardness correlated with high coercivity, mechanical softness with low coercivity (steels were harder than pure iron, and also had higher coercivities). Although this correlation between mechanical and magnetic properties does not extend to many modern materials, the terms "hard" and "soft" have been used for so many years that they're probably here to stay. In fact, they're convenient enough that I'll often be using them from now on myself.

When a temporary (soft) magnet, say a paper clip, is temporarily unmagnetized, the magnetic domains inside it look something like those in Figure 6.1C. (Instead of just four domains, there may be hundreds or even thousands. The idea's the same, but it was easier to draw just four.) Now suppose you bring up a permanent (hard) magnet. Like tiny compass needles, each atomic magnet in the paper clip attempts to align itself with the magnetic field from the permanent magnet. But the atomic magnets can't turn separately, because each is strongly coupled to its neighbors via the exchange force.

Let's suppose the magnetic field is directed upward in Figure 6.1. The atomic magnets in half of the domains are already aligned with the field, but those in the other domains point in the opposite direction. The "aligned" domains will grow, the "unaligned" ones will shrink. If the field pointing up gains strength (if the permanent magnet is brought closer to the object, for example), the domain picture will go from C to D to E. Your paper clip is now magnetized. It won't help to increase the field any further once you've reached E. Once you've got all the atomic magnets lined up, the magnetization of your paper clip—your soft magnet—is "saturated."

Take your permanent magnet away and the reverse happens. Domains reappear, and the picture inside the soft magnet moves from E to D to C. Unmagnetized again. After all, it's only a temporary magnet! Bring up the opposite pole of the permanent magnet and the domains go from C to B to A, and the soft magnet becomes saturated in the opposite direction.

Figure 6.1 shows what's going on inside the soft magnet: magnetic domains are growing and shrinking—by the motion of domain walls. The result, however, can easily be measured from outside the magnet with instruments that measure the *net* magnetization—the difference between how many atomic magnets are pointing up and how many are pointing down. This difference increased from zero at C to a maximum in one direction at E or a maximum in the opposite direction at A. This net magnetization, plotted vertically in Figure 6.2, depends on the strength of the magnetic field you've applied, plotted horizontally. The net magnetization varies rapidly with small applied fields, and the temporary magnet becomes saturated when you've lined up all the atomic magnets. This graph is a somewhat idealized "magnetization curve" of a soft magnetic material.

Many magnetic domains grow or shrink inside your refrigerator door or your paper clip when it becomes temporarily magnetized by a permanent magnet. You can't see the domains, but believe me, it's happening. In my years with General Electric, I spent many hours staring through microscopes at magnetic domains in various materials, watching them grow or shrink when

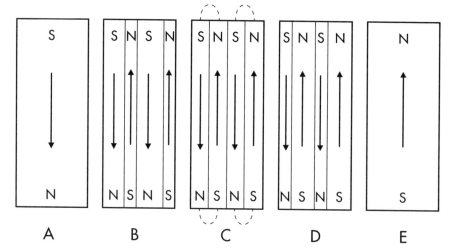

Figure 6.1 A simplified view of magnetic domain structures inside an iron bar. (A) In a saturated magnet, all the "atomic magnets" are pointing in the same direction. (B) When a magnet becomes partly demagnetized, the atoms in different areas (domains) of the bar begin to align in different directions, which weakens the magnetic field of the entire magnet. (C) A completely demagnetized bar has no net magnetization, because the fields of the adjacent but opposite domains cancel each other out. As the bar is remagnetized, this time in the opposite direction, the atoms in the different domains realign gradually so the bar is first (D) partly magnetized and then (E) saturated in the opposite direction. The external magnetic fields in a demagnetized material *(dashed lines in C)* are much smaller than the field outside a saturated magnet (as shown in Figure 1.1A).

I applied magnetic fields. Magnetic domains can be made visible with various experimental techniques (an example is shown in Figure 6.3).

What makes most soft magnetic materials so remarkable is that it takes only a few gauss of applied magnetic field to produce thousands of gauss of net magnetization. Thus a soft magnetic core can greatly amplify the field produced by the current-carrying coil of an electromagnet (fact 7). This amplification factor we call *permeability*. The permeability can be very large because most of the work of aligning the atomic magnets has already been done by the exchange force between neighboring atoms. Within each domain, the material is already magnetized, but in different directions in different domains. All the applied field has to do is to favor some domains over the others. In 1600, Gilbert argued that magnetism in iron "lies confined and asleep, is awakened by a lodestone, . . . and comes into sympathy with it." Looking through a microscope at domain walls moving "in sympathy" with applied fields, we might say the same today.

When domains grow or shrink, all the action is really taking place inside those moving domain walls, where the atomic magnets are flipping over from one direction to the other. (When you zip up your jacket, all the action is taking place inside the zipper. It's easier to engage a few teeth at a time than to engage them all simultaneously.) How easily any magnetic material can be magnetized, demagnetized, or magnetized in the opposite direction ("reversed") depends on how easily domain walls can move.

In the succeeding chapters we'll come across many applica-

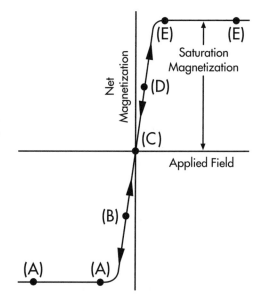

Figure 6.2 Magnetization curve of an ideal soft (temporary) magnet. The net magnetization varies with the strength of the magnetic field applied to it. Letters (A–E) correspond to the domain structure of the material as shown in Figure 6.1. Note that when a temporary magnet is not in the presence of an applied field, it loses its magnetization (C).

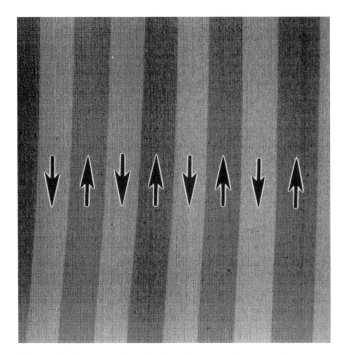

Figure 6.3 Magnetic domains in a demagnetized amorphous-metal ribbon, with arrows added to show the direction of magnetization in each domain. An electron beam traveling from right to left hits the magnet at a glancing angle. The resulting force (fact 9) pushes the electrons into or out of the surface, depending on the direction of the magnetic field (the arrow). Thus one set of domains (the domains where the electrons are pushed out of the surface) scatters back more electrons than the other, and the difference in electron scattering was detected as photographic contrast. Each domain is about one millimeter wide.

tions for which we want permanent magnets, and many others (particularly AC applications) for which we want temporary magnets. If your job is to produce a good soft magnetic material for an electromagnet core or other application requiring a temporary magnet, you want to produce the material in such a way that magnetic domain walls form and move easily. But if your job instead is to produce a good hard magnetic material for a permanent magnet, your goal is just the opposite. You want to produce the material so that it is very difficult to form and move domain walls. Making it easier or harder to move domain walls is, in a nutshell, the goal of magnetic materials science. For many years, that was the goal of my research at GE.

///// Harder and Softer

It's easier to drive your car down a smooth interstate highway than down a bumpy, rock-strewn road. The scientist developing soft magnetic materials wants to create a homogeneous microstructure that provides a smooth "highway" for easy motion of domain walls, with few bumps along the way. The scientist developing hard magnetic materials instead wants to put lots of bumps and rocks in the road.

Most metals and ceramics are crystalline, with the atoms arranged in regular rows and columns like soldiers on parade. But they usually are polycrystalline, containing many small crystals separated by boundaries in which the atoms are more loosely arranged. The average size of the individual crystals in magnetic materials can be as large as centimeters or as small as nanometers. The smaller the crystal size, the more crystal boundaries a given magnet will contain.

In the boundaries between crystals, the atoms are farther apart from their neighbors, and experience weaker exchange forces, than the atoms within the crystals. Any impurity atoms that are present in the material tend to be located in the crystal boundaries. For both these reasons, ferromagnetism is usually a bit weaker inside the crystal boundaries than inside the crystals. For a magnetic domain wall trying to move through a magnet, a crystal boundary can act like a bump on the road. One common way to control the mobility of domain walls is to control the size of crystals, and thereby control the number of crystal boundaries.

One approach to making good soft magnetic materials is to choose a processing method that creates large crystals. Some high-quality silicon steels used in transformers contain crystals nearly a centimeter in size. An alternate approach is to produce an amorphous, noncrystalline material (Chapter 8). If you have no crystals, you have no crystal boundaries. Either approach gives you few crystal boundaries, making it easier for domain walls to move (fewer bumps in the road).

To make a good hard material, you use the opposite approach. You want very small crystals, so that you have lots of crystal boundaries. Crystal sizes in most permanent magnets range from as small as 20 or 30 nanometers up to a few micrometers. You may even want to put a few "rocks" in the road by introducing tiny regions of different chemical composition and, therefore, different magnetic properties—creating an inhomogeneous microstructure somewhat like maple walnut or "rocky road" ice cream. Nature did that in making lodestones, as discussed in Chapter 2, by distributing small regions of maghemite, and regions containing different chemical elements, within the magnetite crystals. That inhomogeneous microstructure made it difficult for domain walls to move through the magnetite, and made the lodestone a permanent (hard) magnet. Modern materials scientists used the same strategy to produce permanent magnets far stronger than lodestones.

With lots of crystal boundaries and chemical inhomogeneities inside the magnet to block domain-wall motion, the magnetization curve no longer looks like Figure 6.2 but instead looks more like Figure 6.4. After you've applied and then removed a large magnetic field pointing upward to the material, it remains magnetized in that direction (a "permanent" magnet). It takes a large magnetic field pointing downward to move domain walls, over all those bumps and rocks, and eventually to reverse the net magnetization. The *coercivity*, first defined in Chapter 2, is high. Remove that downward field, and the magnet remains magnetized downward. (In zero applied field, the material has a bistable memory used to store bits for magnetic recording; see Chapter 9.) It now takes an upward field greater than the coercivity to

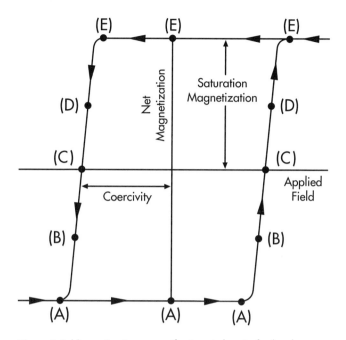

Figure 6.4 Magnetization curve (hysteresis loop) of a hard (permanent) magnet. Letters (A–E) correspond to the domain structures as shown in Figure 6.1. Note that when the applied field is zero, the magnetization is not zero, as it is in an ideal soft magnet (Figure 6.2); instead, a net magnetization remains. This remanent magnetization is either up or down, depending on the direction of the prior applied field—or, put another way, on the *history* of the field. Such behavior is called *hysteresis* (pronounced "history-sis," a convenient mnemonic even though *history* and *hysteresis* have different etymological origins.) The strength of the applied field required to reduce the net magnetization to zero (that is, to demagnetize the magnet) is the coercivity, as explained in Chapter 2.

reverse the magnetization again. A magnetization curve like that in Figure 6.4 is called a "hysteresis loop." (The magnetization curve in Figure 6.2 is for an ideal soft magnetic material, with zero coercivity—absolutely no obstruction to domain-wall motion. Real soft magnetic materials have very narrow hysteresis loops, with coercivities that are very low but not quite zero.)

We can get soft magnetic behavior (Figure 6.2, low coercivity) by minimizing the crystal boundaries and inhomogeneities that obstruct domain-wall motion. We can get hard magnetic behavior (Figure 6.4, high coercivity) by maximizing those obstructions. But how easily domain walls move depends not only on crystal size and microstructure, but also on fundamental properties of the domain walls themselves. (At the risk of overdoing my analogy, I remind you that a jeep is better at handling rough terrain than a low-slung sports car.)

MORE MAGNETIC TERMINOLOGY

hard magnets   Permanent magnets; high-coercivity materials.

soft magnets   Temporary magnets; low-coercivity materials.

Curie temperature   Temperature beyond which magnetic materials lose their magnetism.

permeability   Factor by which the magnetic field of the current-carrying coil of an electromagnet is amplified by a soft-magnetic core. A common quality index for soft magnets, permeability usually varies inversely with coercivity.

magnetic domain   Region within a ferromagnet in which all the atomic magnets point in the same direction.

domain wall   Boundary between two magnetic domains magnetized in different directions. The movement of domain walls is easy in soft magnets, hard in hard magnets.

magnetic anisotropy   Variation in magnetic properties due to differences in the direction of an applied magnetic field. Crystal structures with high anisotropies are potentially good hard magnets, those with low anisotropies are potentially good soft magnets.

microstructure   Arrangement of atoms within a magnet (size and orientation of crystals, chemical inhomogeneities, etc.). Microstructure is determined by processing; its effect on domain-wall mobility controls "structure-sensitive" properties like coercivity and permeability.

The ease with which a domain wall can handle the bumps along the road depends on a fundamental property of the magnetic material called its *magnetic anisotropy*. Some properties of materials are the same in all directions. We say they are *isotropic*. Some properties are different in different directions—they are *anisotropic*. As you probably have noticed, wood breaks more easily along the grain than across the grain. The fracture strength of wood is very anisotropic. Most solids are crystalline, with atoms arranged in regular rows and columns, as noted before. As a result, many properties of crystalline materials are different in different directions—are anisotropic—including their magnetic properties. The degree to which the magnetic properties of a crystal depend on the direction of magnetization is called its magnetic anisotropy. Some magnetic materials are very anisotropic, some are not.

Within a magnetic domain, the atomic magnets prefer to point along certain crystalline directions, directions favored by the material's magnetic anisotropy. Across a magnetic domain wall, however, the direction of the atomic magnets gradually swings from the direction of the magnetization in the domain on one side of the wall to the direction on the other side of the wall. Within the wall, therefore, the atomic magnets are pointing in directions *not* favored by the material's magnetic anisotropy. In materials where the preference for one crystal direction is very strong (high anisotropy), the domain walls are very narrow (only a few atoms thick) and only a few atomic magnets point in "wrong" directions. In materials where the atomic magnets are less picky (low anisotropy), the domain walls are very wide (several thousand atoms thick), with the atomic magnets only very gradually changing direction. Wide walls are not as easily obstructed by narrow crystal boundaries and other small obstacles as narrow walls are. (A broad wheel base helps a truck traverse bumpy roads.)

Low magnetic anisotropy yields wide domain walls and promotes soft magnetic behavior, while high magnetic anisotropy yields narrow domain walls and promotes hard magnetic behavior. So, in addition to controlling crystal size and microstructural inhomogeneities, scientists searching for good soft and hard magnetic materials search for crystal structures with very low or very high anisotropies. They learned, for example, that rare-earth elements promote high magnetic anisotropy, which led to the development of rare-earth permanent magnets like the current champion, neodymium-iron-boron. "Neo" magnets have narrow domain walls that are easily obstructed. In contrast, pure iron is not very anisotropic and has much wider domain walls. Nickel additions lower the magnetic anisotropy further, an observation that led to the development of iron-nickel alloys (permalloys), which are among the best soft magnetic materials. Permalloys have low anisotropies and very wide domain walls—they are less vulnerable to bumps or rocks in the road.

The main factors that control coercivity, the basic measure of magnetic "hardness," are thus magnetic anisotropy, crystal size, and chemical homogeneity. For optimum magnetic properties, the processing of the material should also encourage the many crystals within the magnet to be oriented so that the crystal directions favored by magnetic anisotropy are all nearly parallel to each other. Controlling crystal orientations to produce alignment of preferred magnetization directions is desirable for both hard and soft magnetic materials. Several different processing techniques have been used to achieve this goal.

By searching for materials with very high or very low magnetic anisotropies, and by controlling crystal size, chemical inhomogeneities, and crystal orientations with judicious processing techniques, materials scientists have produced in recent years both soft and hard magnetic materials that are far superior to the materials available a few decades ago. (Advances in hard magnetic materials were discussed in Chapter 4. Advances in soft magnetic materials, particularly amorphous alloys, are discussed in Chapter 8.) Today a wide range of coercivities in soft and hard

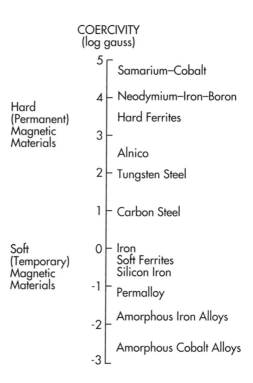

Figure 6.5 The coercivity—the strength of an applied field needed to demagnetize a magnetic material—of various hard and soft magnets, on a logarithmic scale. Amorphous metals (see Chapter 8) may be demagnetized by a field of a milligauss (10^{-3} gauss), whereas rare-earth permanent magnets (see Chapter 4) require fields of tens of kilogauss (10^4 gauss and more) to be demagnetized.

COERCIVITY
(log gauss)

Hard (Permanent) Magnetic Materials

Soft (Temporary) Magnetic Materials

5 — Samarium–Cobalt

4 — Neodymium–Iron–Boron
Hard Ferrites

3 —
Alnico

2 — Tungsten Steel

1 — Carbon Steel

0 — Iron
Soft Ferrites
Silicon Iron

-1 — Permalloy

-2 — Amorphous Iron Alloys

Amorphous Cobalt Alloys

-3

magnetic materials is available (Figure 6.5). Research continues on both ends of the coercivity spectrum, and improvements continue to be made. The end is not yet in sight.

Early in this chapter, I asked a few "simple" questions about soft and hard magnetic materials. To answer those questions even qualitatively, I've discussed spin, exchange, magnetic domains, domain walls, crystal boundaries, chemical inhomogeneities, and magnetic anisotropy. In the first century B.C., the Latin poet-philosopher Lucretius wrote his masterpiece *De Rerum Natura* (On the Nature of Things). Introducing his discussion of the magnetic properties of lodestones, he wrote, "In matters of this sort many principles have to be established before you can give a reason for the thing itself, and you must approach by exceedingly long and roundabout ways." Over two thousand years have passed, and that's still true.

///// Hard Superconductors

Some properties of materials depend primarily on the overall chemical composition and are not very sensitive to microstructural details like crystal size, chemical inhomogeneities, or departures from crystal perfection (atoms out of line). Among magnetic properties, examples are Curie temperature, saturation magnetization, and magnetic anisotropy. But, as we've learned, coercivity is very different. Because it depends on the ease of domain-wall motion, it is extremely sensitive to microstructure. The coercivity (magnetic hardness) of a particular magnetic alloy may increase by a factor of a thousand or more by changes in microstructure. To a materials scientist, coercivity is *structure-sensitive*.

For both soft and hard magnetic materials, we want a high Curie temperature and a high saturation magnetization. A high magnetic anisotropy indicates a *potential* for magnetically hard properties, but the material will have the desired high coercivity only if it can be processed to produce a microstructure that obstructs domain-wall motion. Another important property for hard magnets, discussed in Chapter 4, is *energy product*. That depends on the area within the hysteresis loop (Figure 6.4), which depends on both the saturation magnetization (the half-height of the loop) and the coercivity (the half-width of the loop). Since coercivity is structure-sensitive, so is energy product.

On the other end of the magnetic-materials spectrum, a low magnetic anisotropy indicates a *potential* for magnetically soft properties, but the material will have the desired low coercivity only if it can be processed to produce a microstructure that facilitates domain-wall motion. Easy domain-wall motion also yields the desired high *permeability*. For a given material, we usually find that the permeability varies inversely with the coercivity; halve the coercivity and you will double the permeability. Since coercivity is structure-sensitive, so is permeability.

Thus for both hard and soft magnets, important technological properties are structure-sensitive. How about superconductors? The important technological properties here are critical temperature, critical field, and critical current, as discussed in the preceding chapter. Critical temperature and critical field are not very sensitive to microstructure, but experiments on high-field superconductors done in the 1960s showed that critical current is. Much like the coercivity of magnets, the critical current of superconductors can be increased by a factor of a thousand or more by changes in microstructure.

The superconducting elements (mercury, lead, tin) studied by Kamerlingh Onnes and other low-temperature physicists in the early days had very low critical fields. Magnetic fields up to a few hundred gauss were completely excluded by these superconductors, which we now call "type I" superconductors. (They were not only super-conductors, they were super-diamagnets.) But once the applied magnetic field exceeded a few hundred gauss, these materials lost all their superconductivity.

Not so the high-field superconductors developed in the 1960s, which, as you might already have guessed, we call "type II" superconductors. Instead of completely excluding magnetic field, they allowed partial field penetration in the form of individual quantized "fluxoids" (Figure 6.6). Not as stubborn in resisting magnetic fields as type I superconductors, the more compliant type II superconductors were able, through their compliance, to retain some superconductivity to much higher magnetic fields.

To be useful as wires for high-field superconducting magnets, type II superconductors must be able to carry, without electrical resistance, large currents in the presence of large magnetic fields. In an electromagnet, the current flow is perpendicular to the direction of the magnetic field, hence to the direction of the

Figure 6.6 The magnetic flux lattice inside a type II superconductor, revealed by small magnetic particles that accumulate at each quantized fluxoid. The magnetic field is perpendicular to the figure, and the fluxoids are about 400 nanometers apart.

fluxoids. But current flow perpendicular to magnetic field produces a force perpendicular to both current and field (fact 9). Thus current flow perpendicular to the fluxoids produces a perpendicular force on the fluxoids. If fluxoids move in response to that force, that motion produces heating and electrical resistance, critical current will be low, and the material will be useless for superconducting magnet wire.

But suppose the material has lots of crystal boundaries, chemical inhomogeneities, and other crystal defects. If microstructural features can obstruct magnetic domain-wall motion and produce hard magnets with high coercivities, perhaps they can obstruct fluxoid motion and produce "hard superconductors" with high critical currents. And they do! Scientists learned in the 1960s and 1970s how to produce niobium-titanium, niobium-tin, and other type II (high-field) superconductors with inhomogeneous microstructures that were capable of blocking fluxoid motion, thereby yielding high critical currents. As a result of the production of "hard" superconductors through microstructure control, we now have MRI magnets, the Tevatron, and many other applications of high-field superconducting magnets.

The superconducting oxide ceramics discovered in the 1980s

have critical temperatures and critical fields much higher than niobium-titanium and niobium-tin. They therefore are believed to have the potential of eventually being used for windings of superconducting magnets producing higher fields than present-day magnets, or producing similar fields at higher temperatures. (If MRI magnets required only liquid nitrogen to cool, rather than liquid helium, they would be cheaper to operate.) But to achieve this potential will require the production of microstructures that can better obstruct fluxoid motion, and thereby yield higher critical currents at high fields. The superconducting "hardness" of these materials is not yet enough to produce high-field electromagnets. Several properties of these new materials—such as a weakening of superconductivity at the crystal boundaries, known as the "weak link" problem—make this a very difficult challenge. But significant progress has been made, and efforts continue. Unfortunately, for high-temperature superconductors, it's apparently not easy to be hard.

///// ATTRACTORS, MOVERS, AND SHAKERS

7

///// Using the Force

Permanent magnets and electromagnets exert attractive and repulsive forces on other permanent magnets and electromagnets, and they exert attractive forces on iron and other soft magnetic materials. Magnets also exert forces on moving electric charges (fact 9—Chapter 5). Magnetic fields that vary with time exert a force on electric charges in conductors, thereby inducing electric currents (fact 8). In the succeeding chapters, we'll consider some of the many ways these various aspects of magnetic forces have been used in modern technology.

Let's start at home. In addition to refrigerator magnets and magnets in toys, a typical home has dozens of hidden magnets—like those involved in playing a VCR mentioned in Chapter 1. Magnets store information in audio and video tapes, but we'll get to them, and those in the recording and pickup heads, in Chapter 9. Magnets work hard to produce the picture on your TV screen; we'll discuss them, and the magnets in the power system that brings electricity to your home, in Chapter 8. In this chapter, we'll limit ourselves to magnets used simply to provide an attractive holding or pulling force ("attractors," like those on your refrigerator), those that provide rotational forces ("movers," like those in the motor driving the tape in the VCR), and those that provide vibrational forces to generate sound ("shakers," like those in your TV's speaker).

The magnets holding notes to your refrigerator are usually made of hard ferrites—inexpensive oxides of iron and another metal such as barium. Often the ferrite is in powder form and dispersed within plastic or rubber. That lowers the cost. As cheap as ferrite magnets are, plastics and rubbers are even cheaper.

A recent article in *Newsweek* reported that refrigerator magnets are catching on as collectibles, as "kitsch in the kitchen." Marlou's collection, you'll recall, even had a showing in a New York art gallery. The 1992 *Encyclopedia of Pop Culture* by Jane and Michael Stern has a section on refrigerator magnets. At the recently opened Mall of America near Minneapolis, one entrepreneur did so well selling refrigerator magnets from a cart that she has expanded to a storefront. A magnet store that opened in 1985 on Fisherman's Wharf in San Francisco has expanded to a chain of five stores, including one at Caesar's Palace in Las Vegas. Several dozen stores across the United States are now devoted exclusively to selling "decorative" refrigerator magnets, although most magnets are still sold at general retail outlets. The stores are supplied by over a hundred magnet manufacturers.

Popular designs for decorative magnets include all varieties of foods, household appliances, animals, flowers, miniature reproductions of famous paintings (a popular item in museum gift shops), and celebrities from Marilyn Monroe and the Beatles to Bill and Hillary Clinton. A magnetic version of Michelangelo's *David*, "Dress Me Up David," is available with a complete magnetic wardrobe. Despite their name, many refrigerator magnets are not purchased to hold up notes in the kitchen. One collector was quoted recently in the *Boston Globe* to say, "The worst thing is for a serious magnet collector to use a magnet to hold something up." And not all kitchen magnets are even targeted for the refrigerator. One practical category is dishwasher magnets, which have a door that can be moved from side to side to remind you whether the dishes inside are Clean or Dirty. For do-it-yourself types, *Crafts 'n Things* magazine featured a series of articles on how to make your own refrigerator magnets, which they called "fridgies."

Promotional refrigerator magnets have become a hot marketing device. There are magnetic business cards, calendars, memo boards, sports schedules, picture frames, rulers, key rings, and business logos of all shapes and sizes, including tooth-shaped magnets to advertise dentists and Band-Aid-shaped magnets to advertise hospitals. Free magnets carrying advertising have been included as premiums in boxes of Ritz crackers, cans of Maxwell House coffee, Domino's pizzas, and numerous other products. One Missouri-based manufacturer of promotional magnets has twice been recognized by *Inc.* magazine as one of America's 500 fastest-growing companies. In addition to kitchen magnets, such companies also manufacture magnetic signs to attach to the doors of cars or trucks.

Why do magnets stick to refrigerators? The steel of the refrigerator wall is normally unmagnetized, but it contains many fine magnetic domains of alternating polarity, as in Figure 6.1C. If you bring a permanent magnet near the refrigerator, the field from the magnet causes domain walls in the refrigerator wall to move. Some domains will grow, some will shrink, as in Figure 6.1, and as a result that part of the refrigerator closest to the magnet becomes magnetized. If the north pole of the magnet is the one facing the refrigerator, the domains that grow will be the ones whose south poles are facing the magnet; this change produces a local south pole on the refrigerator and an attractive force that will hold the magnet, and some intervening note paper, in place.

The magnetic field from a permanent magnet is strongest at the poles of the magnet and decreases rapidly with increasing distance from the magnet. If the item you are trying to hold between the magnet and refrigerator is too thick, the field from the magnet that reaches the refrigerator will be too weak to magnetize the refrigerator and produce a strong holding force.

Refrigerator magnets vary greatly in their strength and the distance to which they produce substantial magnetic fields. Some decorative designs, particularly heavy three-dimensional items (such as refrigerators and other appliances with doors that open), are mounted on ferrite disks up to an inch in diameter, magnetized so that one face is a north pole and the opposite face is a south pole. Such magnets should not be placed in close proximity to credit cards, recording tapes, or floppy disks. Most decorative and promotional magnets, however, are mounted on thin plastic or rubber sheets containing ferrite powder. These are magnetized in a pattern designed to assure that the magnetic fields do not extend very far. Instead of just one north pole and one south pole, they are magnetized with many stripes of alternating polarity, each only a few millimeters wide. Placed against a refrigerator, these poles will induce stripes of opposite poles in the refrigerator wall and only a modest attractive force in the area of these stripes. Since the magnetic field from such magnets does not extend very far, they cannot hold up a thick wad of paper, but neither can they damage your credit cards.

You probably have other magnets around your home providing simple holding or pulling forces. For example, the rubber gasket around a refrigerator door is usually magnetic. (Check it out with a paper clip. There are also magnets *inside* a refrigerator!) Safety considerations require the attractive force in the gasket to be rather weak, so that a small child trapped inside an abandoned refrigerator can easily escape. Kitchen cabinets and other swinging doors often have pairs of permanent and temporary magnets to hold them closed. Some garages or basements have strips of permanent magnets mounted on the wall to hold screwdrivers and other steel tools. (And some screwdrivers are weakly magnetized so that they can hold steel screws.) Electric can openers usually have a permanent magnet that holds the top of the can after it's been removed. (Because "tin cans" are attracted to magnets, one of my daughter's elementary-school teachers proved to her class that tin is magnetic. NOT! What many people call "tin cans" are actually *steel* cans with a thin coating of tin for protection against corrosion. They are *tinned* cans, not tin cans.)

In my own home, each window has a small permanent magnet that plays a key role in a burglar alarm system. When the window is closed, the magnet attracts a steel bar in a small electrical device mounted in the window frame. If the window is opened at 3 A.M. by a burglar, the magnet can no longer attract the steel bar, which springs back and closes a circuit that activates the alarm.

One of my daughters installed a cat door to allow her cat to leave and enter the house at any time. He was a very sociable cat, however, and soon many of his alleyway friends were dropping in at all hours of day and night for snacks and mischief. The problem was solved with an ingenious device—a cat door with an electromagnetic latch that opens from outside only when a strong permanent magnet is near it. The cat wears such a magnet on his collar, and the magnet unlatches the door when he approaches. Once he enters, the door re-latches, and any other cats (or squirrels, raccoons, etc.), without magnets, can't get in. The system works well, and probably will continue to do so until the neighbors start putting magnets on *their* cats' collars.

Want to climb walls and ceilings like Spiderman? Now you can—with the "Gripper" (Figure 7.1)—provided that the walls and ceiling are made of steel. Developed by Los Alamos National Laboratory, each Gripper weighs only 1.5 pounds. With one on each hand and foot, you can scale bridges, towers, storage tanks,

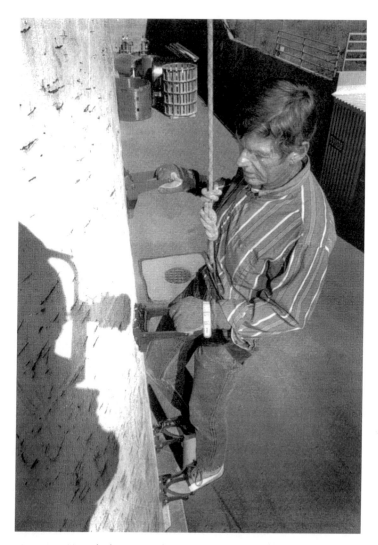

Figure 7.1 Man climbing a steel storage tank with the use of magnetic Grippers on his hands and feet.

offshore oil rigs, or the steel girders of buildings under construction. A hand Gripper provides over 500 pounds of attractive force to a bare steel surface, 200 pounds to a steel surface with a thick coat of paint. Unless you're Hulk Hogan, when you want to detach the magnets for each move up, down, or sideways you'll use the curved cams that lift one side of the magnet from the wall, reducing the force required to dislodge it. (Alternatively, you could bring along a series of Grippers and leave them in place, much as rock climbers leave behind their pitons.) The foot Grippers provide much stronger forces, since leg strength is greater than arm strength. The attractive forces are provided by "neo" (neodymium-iron-boron) permanent magnets, and a steel dowel pin pushes into the surface of the wall to prevent downward slippage. Although considerable commercial interest has developed, the Grippers were originally developed for military applications.

There are dozens of devices in the home, office, or factory that take advantage of the attractive force between a permanent magnet and a temporary magnet. But there are even more devices that utilize the attractive force between an electromagnet and a temporary magnet. As we learned earlier from James Bond and Jaws, an iron-core electromagnet is a very powerful attractor, and one that can conveniently be turned on and off.

At a marina near me, a steel gate at the entrance to the docks is normally held very tightly shut by a strong electromagnet. If you punch in the right numbers on the keypad next to the gate, however, the current to the electromagnet is turned off and the gate can easily be opened. The on-off attractive force achievable with electromagnets has found many applications both large (lifting autos and other heavy steel objects) and small (opening and closing electric circuits). Doorbells, icemakers, printers, automobile horns, and ignitions are just a few common devices utilizing the attractive force between an electromagnet and a temporary magnet.

The Thomas Register is a multi-volume reference work, available in most libraries, that lists industrial products and their suppliers. It names hundreds of suppliers and types of magnets, many of which are for applications based simply on attractive forces, including holding magnets and lifting magnets of many sizes, strengths, and shapes. That didn't surprise me, but I was surprised to see several suppliers of "magnetic sweepers" and "fishing magnets." Magnetic sweepers, like lawnmowers, come in light push models, larger riding ones, and still larger trailer models, to be dragged behind a truck. All are designed to clear factory floors, parking lots, roads, runways, and driveways of steel debris. Fishing magnets are designed to locate and lift anchors, fishing equipment, and other steel items lost underwater, or broken drills, tools, or pipes from the bottom of oil wells. (The same idea applies around the house, when you use a magnet at the end of a string or stick to retrieve keys and other items dropped behind furniture.) There are also many entries in the Thomas Register for magnetic conveyer belts and magnetic separators, the latter designed to separate magnetic and nonmagnetic items (thanks to fact 3). So if you want to hold it, lift it, sweep it, fish for it, convey it, or separate it, you can do it with magnets.

///// Movers

The widest application of magnetic forces is in electric motors. In my home, electric motors move tapes through the VCR, stereo, portable cassette player, and telephone answering machine. They also spin clothes in the washer and dryer, chew up food in the blender and garbage disposal, and compress cooling fluid in the air conditioner and refrigerator (another magnet *inside* a refrigerator). They move air in vacuum cleaners, hair dryers, and fans, and water in the dishwasher. They operate the electric knife, clocks, drill, pencil sharpener, typewriter, printer, garage doors, and the turntable on the phonograph or CD player. And when I click the mouse to access a particular file in my computer, one motor rotates the disk at high speed and another, called a stepper motor, moves the read-write head across the disk very precisely to the appropriate spot.

There are also dozens of electric motors in cars, which is why General Motors was in the magnet business. Figure 7.2 lists some of the possible applications in a modern automobile for the neodymium-iron-boron permanent magnets the company developed. (For most of these applications, however, less expensive ferrite magnets or electromagnets are still used.) Cars rely on magnets not only for many motors, but also for a number of "sensors," like those in the anti-lock brakes. Sensors measure various quantities and deliver appropriate electrical messages to control elements. Magnetic materials are particularly good at sensing force, torque, motion, rotation, and, of course, current and magnetic field. Some of the 34 applications shown in the

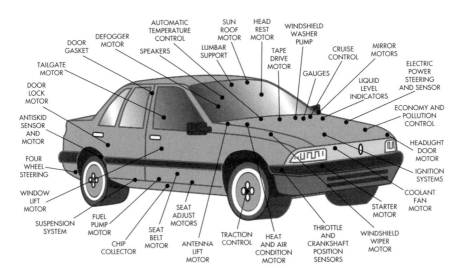

Figure 7.2 Various uses of magnets in automobiles for which neodymium-iron-boron permanent magnets might be used.

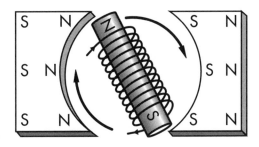

Figure 7.3 Simplified diagram of a DC permanent-magnet motor. The current flowing in the rotor coil magnetizes the temporary-magnet core as shown, and magnetic forces between the poles of the rotor and the stator (the two permanent magnets surrounding the rotor) turn the rotor clockwise. When the unlike poles approach each other, the current to the rotor is reversed, the rotor poles reverse, and magnetic forces continue to turn the rotor clockwise.

figure, like "gauges," refer to multiple devices. (Most of the gauges on the dashboard utilize magnetic forces.) The airbag sensor, not shown in the figure, uses permanent magnets in most cars. And increasingly tough environmental laws, aimed at decreasing air pollution from gasoline engines, will soon be greatly increasing the number of all-electric cars, which will require even more magnets. Motors and other devices using magnetic forces are serving us not only around the house, but also on the road.

Electric motors turn electrical energy into rotational mechanical energy via the attractive and repulsive forces between two magnets. One magnet is stationary and doughnut-shaped, while the other rotates within the hole of the doughnut. They're called, respectively and logically, the stator and the rotor.

The simplest design is the DC permanent-magnet motor (Figure 7.3). Here the stator is a permanent magnet and the rotor is an electromagnet. Attractive forces between unlike poles and repulsive forces between like poles (fact 2) turn the rotor in the direction shown. Just as the rotor is reaching its desired goal (north poles near south poles), however, the current in the rotor is reversed by the commutator, an electrical contact designed to reverse the rotor current every half-turn. This reverses the poles of the rotor (fact 6), at which point two north poles are close together and two south poles are close together. The repulsive force between like poles pushes the rotor around another half-cycle. The current reverses every-half turn, continually frustrating the rotor in its attempt to bring north poles near south poles.

There are many, many variations on this theme. Sometimes the rotor is a permanent magnet and the stator is an electromagnet, sometimes both are electromagnets. Some motors run on DC, some on AC, some on both. Often both rotor and stator are not single magnets but a series of magnets with poles alternating around the circumference. Whatever the details of the design, and whether the motor is a tiny one operating a toy or a giant one in a factory, the forces turning electric motors are essentially the same as in the simple motor in Figure 7.3.

Although most motors turn, turn, turn, some instead convert electrical signals into *linear* motion. "Maglev" trains, for example, are not only levitated but also propelled and braked by magnetic forces between magnets in the track and magnets in the train. The track serves as the stator of a linear motor. Since maglev trains have no contact with the ground while in motion, it is fortunate that magnetic forces can act at a distance (fact 4). More about maglev later.

Japanese engineers are currently designing vertical linear motors to drive elevators in skyscrapers; conventional elevators can't reach beyond about 800 meters, above which elevator cables become too thick and heavy for practical use. Vertical linear motors that drive elevators will, of course, be limited to very moderate accelerations and velocities. Other engineers have proposed using electromagnetic forces to achieve vertical accelerations far higher than our fragile human bodies can withstand, and vertical velocities high enough to escape the earth's gravitational pull. Giant electromagnetic launchers, built along the slopes of hillsides, have been proposed as mechanisms for launching space station components into orbit. Much engineering effort (and much money) has been spent in recent years developing electromagnetic "rail guns" and "coil guns" capable of accelerating objects to speeds of miles per second. The accelerated objects to date, however, have not been potential space station components, but potential Star Wars weapons.

Although electric motors have over a century of history, the recent development of magnetic materials with greatly increased coercivities and energy products (Figure 4.1) has made motor design a rapidly developing field. Fifty years ago, nearly all motors employed only electromagnets. The development of alnico and ferrite magnets led to much more widespread use of permanent-magnet motors, particularly for small, low-power applications. The biggest breakthrough, however, came with the appearance of rare-earth magnets, particularly the introduction of "neo" magnets in 1983. Motors now can be made smaller (with the same power), more powerful (with the same size), and more efficient (because permanent magnets, unlike electromagnets, can provide magnetic fields without energy-wasting electric currents). Many designs are now feasible that were completely impossible if only electromagnets could be used. Many of the latest high-tech devices, including computers and printers, rely on motors and actuators powered by rare-earth magnets. (An "actuator" is like a motor, but it provides limited, rather than continuous, motion.) The increased force per unit mass available with rare-earth magnets has made them the companion components to microelectronics in increasing the speed of doing things.

///// Shakers, Woofers, and Tweeters

Sound is vibration. Whether it's rock, rap, pop, or classical, or the human voice saying "My name is Bond—James Bond," what your ear "hears" is vibrations of the air. And if the sound is coming from speakers in your radio, TV, tape or CD player, from your telephone receiver, headphones, earphones, or from a hearing aid, the vibrations of the air are generated by a piece of plastic, paper, metal, or other material being shaken by magnetic forces.

Although the motion of motors and actuators can be adequately described by the forces between magnetic poles, the forces generating sound in most loudspeakers are more easily described in terms of the forces acting on a current-carrying wire in a magnetic field. For this we'll need one more "fact about the force," the tenth item on the list begun in Chapter 1.

Fact 10: A current-carrying wire in a perpendicular magnetic field experiences a force in a direction perpendicular to both the wire and the field.

Like fact 9, to which this statement is very closely related, there are a lot of perpendiculars in there. Too many for you? Picture a wire lying across this page and carrying current from left to right. If a magnetic field is directed upward from the bottom toward the top of the page, the wire will feel a force pushing it away from the page toward you. If you reverse the direction of either the current or the field, the force will also be reversed—that is, the wire will feel a force pushing it into the page. If you increase the strength of either the current or the field, the force will increase proportionately.

Fact 9 dealt with the force exerted by a magnetic field on a charged particle moving *in free space.* Fact 10 deals with the force exerted by a magnetic field on charged particles moving *in a wire* (electric current). In both cases, the force is perpendicular to both the field and the motion of the charged particles. Charged particles moving in free space respond by moving in circular or helical paths (Figure 5.1). Charged particles moving in a wire don't have that freedom—they must keep moving along the wire. So the force moves the wire and not individual particles. Physicists working with high-energy accelerators or fusion reactors, or studying the ionosphere, are dealing with charged particles in free space and fact 9. Engineers who are designing speakers, motors, or other devices with current-carrying wires in magnetic fields are more interested in fact 10.

I described motor action earlier in terms of fact 2, the attractive and repulsive forces between magnetic poles. It is also possible to explain motors in terms of fact 10 and magnetic forces on current-carrying wires. These are just two ways of describing the

ALL TEN FACTS ABOUT THE FORCE

Known for hundreds of years

1. North poles point north, south poles point south.
2. Like poles repel, unlike poles attract.
3. Magnetic forces attract only magnetic materials.
4. Magnetic forces act at a distance.
5. While magnetized, temporary magnets act like permanent magnets.

Known only since the nineteenth century

6. A coil of wire with an electric current flowing through it becomes a magnet.
7. Putting iron inside a current-carrying coil increases the strength of the electromagnet.
8. A changing magnetic field induces an electric current in a conductor.
9. A charged particle experiences no magnetic force when moving parallel to a magnetic field, but when it is moving perpendicular to the field it experiences a force perpendicular to both the field and the direction of motion.
10. A current-carrying wire in a perpendicular magnetic field experiences a force in a direction perpendicular to both the wire and the field.

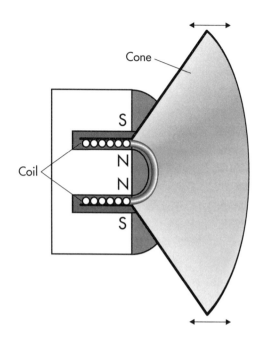

Figure 7.4 Cross-sectional view of a simple loudspeaker; the speaker has circular symmetry. A cylindrical permanent magnet provides a radial magnetic field in a circular gap; this field points from north pole to south pole and so is directed upward in the upper half of the diagram and downward in the lower half. Within the gap a coil of wire carries alternating current. The alternating direction of the current produces alternating forces on the wire (fact 10) that vibrate both the coil and the cone-shaped diaphragm attached to it. The vibration of the cone vibrates the adjacent air and thereby generates sound.

same forces—but fact 10 is handy because it's the easiest way to explain how speakers work.

Most speakers have circular magnets with a gap between north and south poles, within which the magnetic field points radially (radially outward in Figure 7.4, since magnetic field, by convention, points from north poles to south poles). In that gap is a coiled wire carrying a circular current that, as seen in the figure, is always perpendicular to the radial magnetic field. By rule 10, because the current is perpendicular to the field, there is a force on the coil pushing it perpendicular to both field and current, in the direction of the arrows in Figure 7.4. (If that's not clear, you might go back to fact 10 and read down to here again.)

Let's suppose that current in the coil is alternating back and forth 440 times each second. (That alternating current may have been generated from a microphone, from a pickup head, or from some other source—let's not worry about that.) The alternating current will produce a magnetic force that will alternate in direction 440 cycles per second, shaking the coil, and anything attached to it, back and forth at 440 cycles per second. The coil is often attached to a cone-shaped diaphragm (Figure 7.4), and the shaking diaphragm gets the neighboring air vibrating at 440 cycles per second. Waves of vibrating air travel across the room and shake little bones in your ear 440 times each second. You identify this constant tone, if you have perfect pitch, as the middle A on the piano keyboard. (That's the note that the members of a symphony orchestra tune their instruments to just before the concert starts.)

It's useful here to introduce another unit—one that's commonly used when discussing frequencies of vibration, and one that we'll use again in later chapters. The unit is the Hertz, abbreviated Hz, named after German physicist Heinrich Hertz (1857–1894). It is used in place of "cycles per second." The frequency of middle A is thus 440 Hz.

If you're young and healthy, your ear can probably hear sounds within the range of about 20 Hz to 20,000 Hz (20 kHz). Most music uses a large portion of this audible range of frequencies, and an ideal loudspeaker would faithfully reproduce frequencies across this entire range. Unfortunately, no single speaker can do that. So this frequency range is usually divided into parts, with one speaker, called a "woofer," designed to reproduce low frequencies (up to perhaps 1 or 2 kHz), and another, called a "tweeter," designed for higher frequencies. Fancier sound systems add a third midrange speaker, and some even have "subwoofers" and "supertweeters" for the lowest and highest frequencies. In contrast, telephone receivers, designed primarily for reproduction of the human voice, typically are designed only for an approximate frequency range of 350 to 3,500 Hz. That's usually enough to recognize the voice of a friend, but music over the phone is pretty lo-fi.

Many children go through a phase, perhaps near the age of 12, when they take things apart to see how they work. I reached this phase about 50 years late, and, stimulated by writing this book, I have been taking things apart to find the UFOs (unseen force objects)—the hidden magnets. I've recently dissected sound-producing devices ranging from a large speaker, with a moving coil 3 inches in diameter and a cone diaphragm flaring to 11 inches, to a tiny earphone only a quarter of an inch wide. In the earphone, the permanent magnet was ring-shaped, the current-carrying coil was stationary, and the vibrating part (with plastic diaphragm attached) was a thin, circular sheet of steel shaken by varying magnetic fields from the coil. That's close to the design Alexander Graham Bell used to generate sound in the first telephone receiver in 1876, but permanent magnets were so much weaker then that much larger magnets were needed. Until the development of improved permanent magnets in the 1930s, telephone receivers were several inches long, and they were mounted separately from the mouthpiece, because the permanent magnets used were long (but narrow) horseshoe-shaped steel magnets. (You may be too young to remember these receivers, but you've probably seen them in old movies.)

As with motors, the design of telephone receivers, speakers, and similar products has been greatly advanced by the steady and dramatic progress in permanent-magnet materials noted in Figure 4.1. Alnicos, ferrites, and, most recently, rare-earth magnets have allowed engineers to design and build smaller and more efficient devices to generate sound. Teenagers can now put on tiny earphones enhanced by rare-earth magnets and turn up the power until it Hz. That's progress! (Fortunately for them, rare-earth magnets have also led to improved hearing aids.)

///// AC, RF, TV, AND EAS

8

///// AC/DC

When you switch on a flashlight, you complete an electric circuit, and electrons flow from the negative terminal of the battery, through the tiny filament in the bulb (thereby heating it enough for it to radiate light), and back to the positive terminal of the battery. Current that flows like this, at a constant rate in one direction only, is called DC or direct current. When you switch on lamps or appliances powered from wall sockets in your home, however, the electric current is continually varying. It builds up to a maximum in one direction, decreases to zero, reverses, reaches a maximum in the opposite direction, decreases again to zero, and repeats this cycle many times each second. This is AC or alternating current, provided by electric power companies in the United States at a frequency of 60 Hertz (50 Hz in Europe).

In Chapter 5, I used an analogy between electric current and water flow. You can experience AC water flow if you're swimming near the shore in the alternating push and pull of heavy surf. Tidal currents are also AC, but they complete only 2 cycles per day. That's a frequency of only 23 microHertz (23 millionths of a cycle per second).

Most electrical devices would work fine on DC (although some would require minor modifications). So why do power companies provide our homes with AC? The first power system introduced by Thomas Edison in New York City in 1882 was DC, and he fought for many years to make DC the national standard. But the "battle of the currents," as it was then called, was won by AC, primarily because of an important magnetic device—the *transformer.* Over the course of the campaign, however, many dogs and cats, two calves, one horse, and a man named William Kemmler were electrocuted in support of the DC cause.

The technical issue underlying the AC/DC controversy centered on the problem of transmitting electric power over long distances—from where it is generated to where it is used—without wasting too much energy along the way. To understand this issue, it's important to note that *power* is the energy delivered per second. Electric power (measured in watts) equals the voltage (in volts) times the current (in amperes). Both AC and DC can be transmitted with minimal energy losses if the current is low and the voltage is high—as high as many thousands of volts (many kilovolts). However, such high voltages are very dangerous and must be reduced to much lower values—roughly one or two hundred volts—before being brought into the home. The winning advantage of AC is that voltage can easily be increased or decreased with the use of transformers, thanks to fact 8, the induction of electric currents by changing magnetic fields. With DC, by contrast, currents and fields are constant in time. Without varying fields, there is no electromagnetic induction, and no easy and efficient method of increasing and decreasing voltage.

A transformer is essentially a doughnut of iron, or other *temporary*-magnet material (a permanent magnet wouldn't work here—the fields are always changing!), around which are wound two coils of wire, the primary and the secondary (Figure 8.1). Alternating current flowing through the primary coil creates a magnetic field (fact 6) that magnetizes the iron doughnut, first in

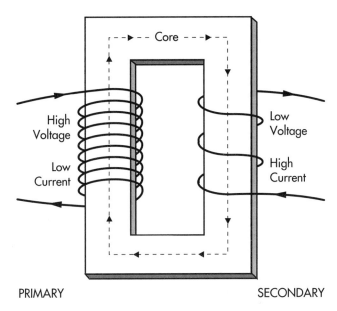

High
Voltage

Low
Current

Core

Low
Voltage

High
Current

PRIMARY

SECONDARY

Figure 8.1 Simplified diagram of a transformer. AC electrical power entering the primary coil at the left generates alternating magnetization of the temporary-magnet core, which induces (fact 8) AC electrical power in the secondary coil at the right. The voltages are proportional to the number of turns in each coil. In this case, the primary voltage is three times the secondary voltage, and the secondary current is three times the primary current. Run this way, this device would be a step-down transformer, like those delivering power to your house.

one direction and then the other, at a frequency of 60 Hz in the U.S. power system. This fluctuating magnetization of the iron creates a changing magnetic field inside the secondary coil, which, by fact 8, induces electric current in that coil. We have used the iron, with the help of facts 6 and 8, to transform electric power in the primary or input coil into electric power in the secondary or output coil.

All we've done so far is move electrical power from one coil to the other. What I haven't mentioned yet is that the voltage induced in the secondary coil depends on how many turns each coil has. If the secondary coil has fewer turns than the primary coil, the output voltage across the secondary is less than the input voltage to the primary. We have done what we were after—we have found a way to change the voltage! A transformer like this, with the output voltage less than the input voltage, is called a step-down transformer. If, instead, the secondary coil had more turns than the primary coil, the secondary voltage would be more than the primary voltage and we would have a step-*up* transformer.

The step-down transformer does not increase or decrease electrical power. Although the output voltage of a step-down transformer is less than the input voltage, the output current is greater than the input current, leaving the product of voltage and current, the power, unchanged (except for power lost to electrical resistance and heating). This transformer merely converts high-voltage, low-current AC power, suitable for long-distance transmission, into low-voltage, high-current AC power, suitable for your home. A step-up transformer does the reverse.

At the power plant where electricity is generated, a large step-up transformer increases the voltage to several hundred kilovolts for efficient long-distance transmission. After traveling many miles, the power arrives at a substation, where step-down transformers lower the voltage to a few kilovolts for safe delivery to your neighborhood. There, a smaller "distribution" transformer (often about the size and shape of a garbage can, easily visible on power poles) lowers the voltage *again* to 120 or 240 volts before delivering it to your house. Thus there are at least three, and usually more, transformers between the generator and your wall socket. They all are made possible by fact 8, the induction of electricity via *changing* magnetic fields, which is impossible with a DC system.

Four years after Edison's first DC system, engineer William Stanley introduced an AC system in Great Barrington, Massachusetts, complete with step-up and step-down transformers and high-voltage transmission. George Westinghouse was impressed, and he soon became the leading proponent of AC power systems. The DC advocates then attempted to undermine the major advantage of AC systems by lobbying legislatures to limit transmission voltages to several hundred volts and initiating a campaign to convince the public that AC was inherently more dangerous than DC.

The leader in publicizing the safety issue was an electrician and inventor named Harold Brown, a DC proponent who accordingly received strong support from Edison and his colleagues. At Columbia College in July 1888, Brown gave a public demonstration of the deadliness of AC by electrocuting a dog with only 300 volts AC, after the dog had survived 1,000 volts DC. After killing many dogs and cats in later "experiments," Brown developed a new strategy. New York State was then seeking a better method than hanging to carry out the death sentence, and Brown suggested alternating current!

Brown stepped up his campaign and killed two calves and a horse to convince authorities that even animals larger than humans could easily be dispatched with AC. The demonstrations were convincing, and New York became the first state to adopt the use of the electric chair—an AC electric chair. Convicted murderer William Kemmler was executed in Auburn State Prison in August 1890, with the use of AC equipment that Brown and supporters had conveniently arranged to be Westinghouse equipment. Competition then ensued to name this new method of execution. It came to be called electrocution, but Edison's supporters suggested "Westinghoused."

Westinghouse and other AC proponents argued against Brown's claims that AC was more dangerous than DC, and Brown responded with a challenge to what amounted to a duel by electricity. Brown would take DC through his body and Westinghouse would take AC, at gradually increasing voltages until one cried "enough" and admitted he was wrong (or died and *proved* it!). Westinghouse declined the offer. (He was wise.

Whether AC is really deadlier than DC, or vice versa, depends in a complicated way on details of the construction of the electric power sources and how the power is delivered to the body, but both currents can do considerable damage.)

Technological controversies nowadays are often strongly influenced by public perception. But in the "battle of the currents," the decision makers paid more attention to the bottom line—the great economic advantage of long-distance high-voltage transmission possible only with AC. Despite Brown's success in branding AC in the press and in the public mind as the "executioner's current," the inherent advantages of AC power ultimately won the battle. Westinghouse was awarded the well-publicized contract to light the 1893 Chicago World's Fair with AC, and in 1895 a giant hydroelectric plant at Niagara Falls began to deliver AC power to Buffalo, twenty miles away. In 1908, Edison said to George Stanley, son of the man who installed the first AC system, "Tell your father I was wrong." By 1917, more than 95 percent of the electric power generated in the United States was AC, thanks to the voltage-changing magnetic magic of transformers. Today AC power is almost universal.

Transformers are not the only components of electric power systems based on magnets and fact 8. Electric *generators* also rely on the use of magnets and electromagnetic induction. Consider the simplified electric motor in Figure 7.3, which produced rotational motion by passing electric current through the electromagnet in the rotor. Instead we can turn the rotor by attaching it to a wheel or turbine that is turned by falling water (as at Niagara Falls) or by expanding steam heated by burning coal, oil, or gas or by a nuclear reactor. Turning within the magnetic field of the stator, the rotor coil experiences changing fields that induce or "generate" electric current. Whereas the electric motor turns electric energy into rotational mechanical energy, the generator does the reverse.

Generators create electrical energy not only at giant power plants but also under the hood of your car. There, a much smaller generator uses the rotation provided by the gasoline engine to keep the car battery charged and to deliver electric current to the lights, heater, and all the electric motors in the car. Everywhere in today's technology, magnets in motors and generators assist us by converting energy from one form into another.

///// Catching the Waves

As a result of Westinghouse's victory over Edison, the electric current that today lights your lamps and heats your toaster is alternating at sixty cycles per second—60 Hz. Think that's pretty fast? The electric currents in the wires that are generating sounds in the speakers in your radio, television, stereo, and telephone are alternating at the frequencies of audible sound, and *audio* frequencies, you'll recall, are up to 20 kilohertz. So those currents are alternating up to twenty thousand times a second!

Think *that's* pretty fast? The currents in your radio and television antennas are alternating at the frequencies of the radio waves that travel through the air from the station's transmitter to your house—at *radio frequencies* (RF). These can be more than 100 megahertz, so currents induced in the antenna can alternate more than a hundred million times a second!

Radio waves consist of wiggling electric and magnetic fields that travel at the speed of light—186,000 miles, or 300,000 kilometers, per second. They're characterized either by their frequency in Hertz or by their wavelength—the distance between one wave crest and the next. The wavelength times the frequency equals the speed of light—a constant—so increasing frequency corresponds to decreasing wavelength. Figure 8.2 shows, on a logarithmic scale, the approximate frequency and wavelength ranges for various types of radio waves and also for other types of electromagnetic waves—ELF (extremely low frequency) 60-Hz power-line frequencies, microwaves, infrared, visible light, ultraviolet, x-rays, and gamma rays.

OK—but what does this all have to do with magnets? Con-

Figure 8.2 The frequencies and wavelengths of various forms of electromagnetic radiation. Note that as frequency increases, wavelength decreases; their product remains constant and equal to the speed of light.

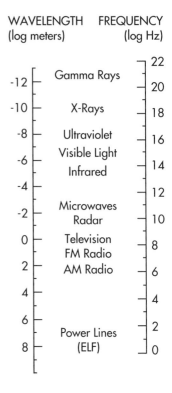

WAVELENGTH (log meters)		FREQUENCY (log Hz)
		22
-12	Gamma Rays	20
-10	X-Rays	18
-8	Ultraviolet	16
-6	Visible Light	14
	Infrared	
-4		12
-2	Microwaves	10
	Radar	
0	Television	8
	FM Radio	
2	AM Radio	6
4		4
6		2
8	Power Lines (ELF)	0

sider the poor radio antenna. Radio waves—wiggling electric and magnetic fields—from dozens of distant radio stations are all simultaneously hitting the antenna, which has the task of translating all that electromagnetic chatter into music, news, or the views of Rush Limbaugh.

The antenna has two serious problems to overcome in bringing you what you *want* to hear. First of all, the electromagnetic fields are very weak, having spread over a huge area after being sent out by each station's transmitter. Second, there are signals from many stations, and you would prefer to listen to one at a time. Fortunately, magnets can help solve both those problems.

Part of the antenna wire is usually coiled around a rod of a magnetic oxide called a *soft ferrite,* thus making it a ferrite-core electromagnet. The magnetic ferrite senses the alternating magnetic field of the radio waves and amplifies it by hundreds or even thousands of times. (The amplification factor, you'll recall from Chapter 6, is called the permeability.) Ferrite rods in radio antennas turn weak signals into much stronger signals. First problem solved.

(Like the hard ferrites discussed in Chapter 4, soft ferrites are compounds of iron oxide with other metallic oxides. The ferrite used in an antenna must be a *temporary*-magnet material, though, because the fields are alternating—as they were in the transformer. Whereas hard ferrites have crystal structures with high magnetic anisotropies, soft ferrites have very different crystal structures, with *low* magnetic anisotropies.)

How about the second problem? The radio receiver must separate the signals of one station from all the rest, a process we call tuning. We rely here on another ferrite-core electromagnet, an *inductor,* in combination with a nonmagnetic *capacitor.*

Recall fact 8 again—changing magnetic fields induce electric fields. Magnetic fields wiggling at a higher frequency are, after all, changing at a faster rate, so the voltage across an inductor increases with increasing frequency. For totally different reasons, the voltage across a capacitor *decreases* with increasing frequency. (No point in trying to explain that—this book is about magnets!) By cleverly combining the opposing behaviors of inductors and capacitors, the tuning circuit responds preferentially to only a very narrow range of frequencies. Fortunately, each station in your neighborhood transmits on a different frequency, so the tuning circuit enables you to hear only one station at a time. (The knob you turn, or the button you push, to move from station to station usually varies the capacitor, not the inductor.)

Now both problems have been solved. A ferrite-core electromagnet, as part of the antenna, amplifies the weak radio signal, and another electromagnet, an inductor in the tuning circuit, helps select a particular frequency corresponding to a particular station.

Another job done by ferrite-core electromagnets, one I mentioned back in Chapter 1, is producing the horizontal and vertical sweeps in the television picture tube. At the back of the tube, a metallic filament is electrically heated to such a high temperature that electrons are "boiled" out of the metal. The electrons are then accelerated by a strong electric field to the screen, where they generate images by exciting light-emitting "phosphors" on the inside surface of the screen. Ferrite-core electromagnets around the neck of the tube provide fast-varying magnetic fields that deflect the electrons in a series of horizontal lines across the screen. By rule 9, the pair of electromagnets that produce the vertical magnetic fields provide the horizontal sweep (those perpendicular forces again!), and the pair producing the horizontal magnetic field provide the vertical sweep.

The television picture tube is the most familiar example of what we call a *cathode-ray tube* (CRT). (Cathodes are negative electrodes, and electron beams emitted from a cathode, in this case the heated filament, are called cathode rays.) The monitor on personal computers is also a CRT, and electromagnets direct the horizontal and vertical sweeps on this screen as well. Other examples of CRTs include radar screens used by air-traffic controllers and oscilloscopes used in laboratories throughout the world.

///// Catching the Crooks

We've discussed the hidden magnets in motors, speakers, transformers, generators, radios, and TVs. These magnets are "hidden" in the sense that they are out of sight and, therefore, to most people, out of mind—but they weren't *intentionally* hidden. Over the last few decades, a new industry has arisen based on a technology in which magnets *are* intentionally hidden. These magnets are the heart of EAS—electronic article surveillance—used in the perennial war against shoplifters.

When I rent videotapes at my local video rental store, I walk out between two large loops of plastic—but I'm not handed the tape until I've passed those loops. When I check books out of the MIT library, the librarian always places the books on a device behind the counter before giving them to me and allowing me to leave—between two large loops of plastic. Something very similar happens when I purchase certain items at my neighborhood drugstore, but here the device the clerk places the items on is in plain sight—and bears the warning: "Please do not place credit, bank, or ATM cards on this pad." That's a clue! As I'll explain in the next chapter, credit cards and bank cards are magnetically coded. Could it be that these mysterious devices used by the librarians and sales clerks, and these mysterious plastic loops that flank the exits of libraries and stores, have something to do with magnets?

Hidden somewhere in the library books, videotapes, and drugstore merchandise are small magnets—sometimes just a strip or wire of a temporary magnet, sometimes a combination of a temporary and a permanent magnet. In the MIT library and my neighborhood drugstore, the magnetic "tags" are deactivated by the librarian or sales clerk before I leave. Otherwise, the AC magnetic fields generated by the antennas within those plastic loops by the exit door would detect a telltale signal from the tags, and an alarm would sound. That keeps us from taking books from the library without checking them out, and from removing items from the drugstore without paying for them.

My video rental store instead has a cheaper type of magnetic tag that can't be deactivated. The clerks there simply let me go through the loop antennas without the tape and then hand the tape to me. One day, while I was checking out a book at the MIT library, a student passed through the exit loops and set off the alarm. Great excitement ensued, but it turned out he wasn't trying to steal library books. He had been to a video rental store before coming to the library—and had a videotape with an active magnetic tag in his briefcase.

The world EAS market, now nearly a billion dollars a year, is growing at more than 20 percent each year. The electronic detection circuits connected to the loop antennas are pretty sophisticated—they must detect small magnetic tags but not give false alarms from other magnetic or metallic objects. The tags must also be designed to give very specific signals when "interrogated" by the loop antennas—a response easily differentiated from the responses of paper clips, pens, or other magnetic items the customer may be carrying. I'll provide no further details about EAS systems in this book—I don't want to give shoplifters any ideas for developing countermeasures!

///// Improving on Iron

In Plato's day, the best hard magnetic materials were lodestones and the best soft magnetic material was iron. Twentieth-century materials science has produced permanent magnets—alnicos, ferrites, rare-earth magnets—far better than lodestones, as we saw in Chapter 4. For AC applications like those discussed in this chapter, materials science has also produced temporary magnets far better than iron.

A good hard magnetic material has a high coercivity; it is hard to demagnetize. Once magnetized, it retains that magnetization (Figure 6.4), making it a "permanent" magnet. A good soft magnetic material has a *low* coercivity; it is *easy* to demagnetize. Once a magnetizing field is removed, it loses its magnetization (Figure 6.2), making it a "temporary" magnet. In devices run by AC, a temporary magnet may be demagnetized and remagnetized many times each second!

The purpose of the temporary-magnet core in an electromagnet is to amplify the magnetic field of the coil (fact 7). You want a material that requires only a few gauss of magnetizing field from the current-carrying coil to move magnetic domain walls (Figure 6.1), and thereby get many gauss of net magnetization from the core. This amplification factor, the permeability, should be as high as possible, and the coercivity as low as possible. For low coercivity and high permeability, we want a material in which magnetic domain walls move easily.

Today we have several materials with much lower coercivities, and much higher permeabilities, than iron (Figure 6.5). For example, the nickel-iron alloys called permalloys (because of their high permeabilities) fill this bill, but they're also more expensive. And today's iron is much better than Plato's, primarily because it's purer. It has a coercivity low enough, and a permeability high enough, for many applications. The reason it's not used in most AC applications today lies in the dark side of fact 8.

Changing magnetic fields produce electric fields—that's the basis of transformer action and the reason Westinghouse and AC beat Edison and DC. But those changing magnetic fields induce electric fields not only in the coils of copper wire, where you want them, but also in the magnetic core itself, where you *don't* want them. That's the dark side of fact 8. If the core is iron, the electric fields produce large "eddy currents" in the core, which waste energy through heating the core material. The heat generated by eddy currents causes undesirable heating in transformers, motors, and other devices and wastes billions of dollars of energy every year.

The major method used to decrease these energy losses has been to add alloying elements to iron to increase its electrical resistance, thereby decreasing the eddy currents. Alloys of iron and silicon replaced pure iron in most 60-Hz transformers many years ago. Permalloys replaced iron in other applications in which their superior properties justified their higher cost. Magnetic alloys are usually used in the form of thin sheets, since this form also limits the eddy currents.

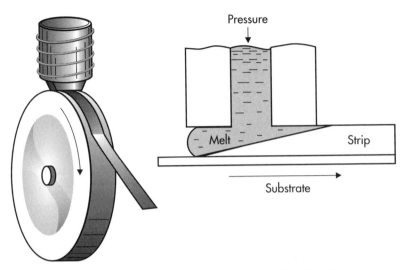

Figure 8.3 (A) An amorphous-metal strip is made by expelling liquid metal onto a rapidly rotating cooled metallic wheel; cooled by the wheel, the liquid metal quickly solidifies into a thin strip. (B) Enlarged cross-sectional view of the solidification process. The liquid metal cools a thousand degrees in a millisecond, too fast for the atoms to form the normal crystalline structure of most solid metals.

Eddy-current energy losses increase very rapidly with increasing frequency, making even these alloys unsuitable for RF applications, such as radio antennas. Here eddy currents are almost completely eliminated by using soft ferrites—magnetic iron oxides that are electrical insulators. Whereas hard ferrites used for permanent magnets are usually barium iron oxide or strontium iron oxide, the soft ferrites usually contain, in addition to iron and oxygen, nickel and zinc or manganese and zinc.

One important approach for producing improved temporary-magnet materials resulted from the recent discovery of an entirely new class of alloys—*amorphous alloys*. As noted earlier, most solids are crystalline materials whose atoms are arranged in regular rows, like soldiers on parade. But some solid materials—window glass, for example—instead have atoms arranged in irregular fashion, like atoms in a liquid and like people in a mob. Such noncrystalline solids are called amorphous.

About thirty-five years ago it was discovered that metals could be made amorphous by adding certain "glass-forming" elements and then cooling the resulting alloy from the liquid state at extremely rapid rates—up to a million degrees per second (about a thousand degrees in a millisecond). Thin sheets of amorphous iron alloys are now made in quantity by squirting molten alloys

Figure 8.4 Examples of amorphous-metal technology from Allied-Signal, Inc.: coils of amorphous metal ribbons (1) and a variety of applications, including transformer cores (2 and 5), electronic article surveillance sensors (3), a computer floppy disk drive (4), power supply (6), bone growth stimulator (7; see Chapter 14), and panel for magnetic shielding (8; see Chapter 15).

onto a rapidly rotating, cooled metal wheel. (See Figures 8.3 and 8.4.) An alloy of iron, silicon, and boron is currently the most used.

The early research on amorphous metals involved tiny "splats," milligrams of metal cooled by propelling liquid drops against cooled metal surfaces. In the original experiments on ribbon production by rapid solidification, a few grams of metal were melted to produce ribbons a few millimeters wide. Research production soon progressed to pounds of metal and ribbons an inch or two wide, and commercial production at Allied-Signal, Inc. today yields amorphous metal strip up to eight inches wide at a rate of two tons an hour!

In transformers built with amorphous alloys, the energy lost to eddy currents is only one-third of the loss in transformers built with the iron-silicon crystalline alloys that were used previously, and less than one-tenth of the loss in the iron-core transformers used a century ago. Many of the distribution transformers being ordered by utility companies in the United States today contain amorphous alloy cores. This technology was pioneered by General Electric (and was based on research done in the materials group in which I worked, the same group that pioneered the development of rare-earth magnets and high-field superconductors). The early development of amorphous-metal production at Allied, and of transformer design and construction at GE, was funded by the Electric Power Research Institute (which in turn is supported by U.S. electric utility companies). For new technologies, transition from the initial low-volume, high-cost stage to the high-volume, low-cost stage is always a challenge. In the case of amorphous-metal transformers, the challenge was successfully met by close cooperation between the materials producer (Allied), the device manufacturer (GE), and the customer (the electric utilities). We all benefit from the net result—less energy wasted to eddy currents, thanks to improved soft magnetic materials.

///// THANKS FOR THE MEMORIES

9

///// Remembering Things Passed

To impress you with the many hidden magnets in today's technology, I listed in Chapter 1 the numerous magnets involved in playing a James Bond movie on your VCR. Since then, I've explained in more detail the magnets in the motor and the speaker (Chapter 7), the magnets sweeping currents across your TV screen (Chapter 8), and the magnets in the generators and transformers that deliver electricity to your home (also Chapter 8). To complete the picture, we need to discuss the electromagnet in the recording head, which converts magnetic fields from the moving tape into electric currents, and the videotape itself, in which many tiny magnets somehow have stored all the sights and sounds of a two-hour movie.

Magnetic recording in all its variations—videotapes and audiotapes (sound only), data storage in computer hard disks and diskettes ("floppies"), credit and ATM cards—is today's largest and fastest-growing use of magnets. The impact of this technology on our lives has been enormous.

It all started back in 1898 with Valdemar Poulsen, a Danish engineer working for the Copenhagen Telephone Company. He apparently wasn't home that often and thought it would be great to have a device that could record telephone messages if someone called when you were out. He therefore invented the "telegraphone," with which you could record the human voice on a wire. He demonstrated it to his friends with a steel piano wire stretched across his laboratory and a trolley holding a little electromagnet (Figure 9.1). He connected a telephone speaker to the electromagnet and yelled into the speaker as the trolley rolled along the wire.

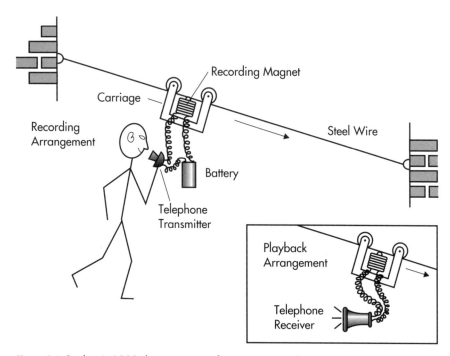

Figure 9.1 Poulsen's 1898 demonstration of magnetic recording on a steel wire. Poulsen spoke into the telephone mouthpiece (transmitter) as the carriage rolled along the wire; currents from the transmitter induced a magnetic pattern—a "memory"—in the wire. When the transmitter was replaced by a receiver and the carriage rolled along the wire again, his voice could be heard in the receiver because the magnetic pattern "recorded" in the wire generated sound from the receiver.

Poulsen's friends were puzzled by his odd behavior. But when he reached the end of the wire, he slid the trolley back to the starting point, replaced the telephone speaker with a telephone receiver, and handed the receiver to a friend. When the friend then slid the trolley down the wire again, he could hear Poulsen's voice faintly in the receiver! Apparently the electromagnet, energized by the currents produced in the speaker by Poulsen's voice, had somehow induced in the steel wire a distinctive magnetic pattern—a "memory." Later, when the same electromagnet (today called a "head") was attached to a receiver, that pattern of magnetic changes generated the same sounds that had produced the pattern.

Poulsen entered his invention in the Paris Exposition of 1900, where it was a great sensation and won the grand prize. It was, however, not a commercial success, because it produced a very weak and distorted sound. The later development of vacuum-tube amplifiers solved the first problem, but unfortunately not the second: amplification of distorted sound just makes the sound harder to ignore. An electronic trick called "AC bias" later improved sound quality, and wire recorders were used by the United States and their allies in World War II. But the Germans apparently had something better. Throughout the war, British and American radio monitors wondered how Hitler's recorded speeches managed to sound so much like live broadcasts.

This mystery was solved in the final days of the war, when U.S. troops stormed the studios of Radio Luxembourg. There they found a new device, a magnetophone, reeling off one of Hitler's last harangues, recorded not on wire but on plastic *tape* coated with many tiny particles of iron oxide. The American commander packed up the machine as booty and shipped it to California, where the newly formed Ampex Corporation soon developed a greatly improved device. Magnetic tape recording hit the headlines in 1947 when it was used to record and broadcast the Bing Crosby radio show. Soon the commercial audiotape

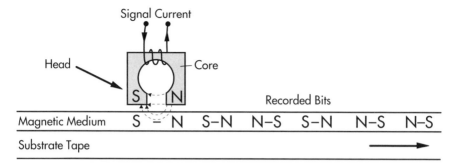

Figure 9.2 Schematic diagram of magnetic recording on a moving tape. Currents into the coil magnetize the temporary-magnet core, the poles of which will reverse if the signal current reverses. The field across the gap then magnetizes the adjacent tape S-N or N-S. The moving tape therefore retains a memory of the current history in the coil.

business was booming. Before too long, videotape recording for television shows became possible, and by the 1970s home VCRs were introduced. (Videotaping is more difficult than audiotaping. Pictures require a lot more magnetic memory than sounds.)

Magnetic recording works because permanent magnets have a memory (Figure 6.4). You can induce north and south poles in a temporary-magnet material like a paper clip by applying a magnetic field, but the poles disappear when you remove the field. The poles of a permanent-magnet material remain where they are. If the field had pointed to the right, you'd leave a north pole on the right end (S-N). If it had pointed to the left, the north pole would be at the left end (N-S). A permanent magnet remembers which way you magnetized it.

In Figure 9.2, a recording tape full of tiny permanent magnets (particles of iron oxide, say) is moving past the head, a "core" of temporary-magnet material with a coil of wire around it and a narrow gap next to the moving tape. Send a pulse of current into the coil and the resulting magnetic field (fact 6) creates an even larger field in the core (fact 7). The field across the gap in the core will then magnetize the part of the tape that is next to it at the time. The tape moves on, but in that piece of it the direction of magnetization (say N-S) stores a memory of the direction of the current pulse. A fraction of a second later, send a pulse of current in the opposite direction into the coil and you'll now magnetize another bit of the tape in the opposite direction (S-N). If you keep doing this, and the tape keeps moving, it will store along its length a memory of the complete history of the current in the coil. That current may have been generated by a single human voice or by the Boston Symphony playing Beethoven's Fifth Symphony. Either way, the head has "written" a magnetic memory of that sound into the tape.

The writing process transformed electric currents into a magnetic memory. You can now use that magnetic memory and electromagnetic induction to re-create those currents. Rewind the tape and run it past the head again, preferably at the same speed. As the various pieces of tape magnetized S-N or N-S pass by, the head is exposed to varying magnetic fields. This induces varying magnetizations in the core and, via fact 8, varying currents in the coil. Send those currents to your earphones, and you'll hear the symphony again. We say that the head is now "reading" what you have previously written into the tape.

The memory stored in your magnetic tape is not necessarily stored forever. The tiny magnets in your tape can lose their memory if they get too hot; don't store your tapes on the radiator. They can also lose their memory if they get exposed to a magnetic field stronger than their coercivity, which is usually a few hundred gauss for a typical tape. This is actually an advantage. You can erase the tape's memory and record something else. However, if you like what you've recorded, you'd better keep it away from large magnetic fields.

In the movie *The Big Easy*, Dennis Quaid played a New Orleans cop who was videotaped accepting some money from a bartender, and the tape was going to be used in court to prove he was on the take. In the next scene, Quaid, in disguise, entered an electrical equipment store and bought a large permanent magnet. (The magnet was identified as an alnico magnet, probably the first and only time in the history of Hollywood that the word *alnico* appeared in a movie script.) Next he was seen throwing the magnet through a bank window, which made the magnet evidence of a crime. The policeman who put the magnet in a police property room where other evidence was stored, apparently being a friend of Quaid's, placed it conveniently atop the videotape. The case against Quaid then had to be dropped, because the taped evidence had been "accidentally" erased. Rodney King is probably very glad that the videotape of his beating reached the local media before it reached the property room of the Los Angeles Police Department.

Exposure to large magnetic fields can be a problem even for the recording material most of us carry in our wallets—the black stripes on the back of credit cards and ATM cards. Credit card numbers, bank account numbers, and other pertinent data are stored magnetically in iron oxide particles in those black stripes.

I was recently invited to speak about magnetic materials at a meeting of a local materials society. I like to enliven my talks with demonstrations, and for that purpose I slipped a powerful rare-earth magnet in my jacket pocket when I left my office for the meeting. My talk, demonstrations included, went well but two days later, when I tried to withdraw some money from the bank with my ATM card, the machine refused me. I had no idea why. With little cash in my pocket, I had to use my credit card for purchases at a department store. I noticed that when the clerk ran my card through the card-reading device, it did not record properly. He accepted the card but had to copy the number down by hand. It was then that I realized that the geometry of men's clothing is such that jacket pockets are not that far from the back pockets in trousers. Two days earlier, the magnet in my jacket pocket had apparently been close enough to the wallet in my back pocket that its field exceeded the coercivity of the tiny magnets in my cards and erased the magnetically stored data. If you have credit and ATM cards in your wallet, keep magnets away!

As consumers, we are most familiar with magnetic information storage in the form of audiotapes and videotapes that store sounds and pictures. But our credit and ATM cards give us a glimpse of a more widespread use of magnetic recording—the storage of numerical data. Government agencies, the military, businesses, scientists, and engineers store vast amounts of data daily in the form of patterns of magnetic poles (S-N, N-S) in those tiny permanent magnets in tapes and disks.

Magnetic recorders in satellites gather data from measuring instruments throughout the orbit, store it, and transmit it when within range of a receiving station. Space probes gather data rapidly during flyby of a planet, store it, and transmit it later at a pace more convenient to earthbound receivers. Back on earth, magnetic data recorders monitor electrocardiograms, radar signals, weather information, industrial processes, auto emissions. They operate typewriters, machine tools, smart weapons, robots—even themselves. They count people, goods, and money, keeping track of national employment figures, supermarket inventories, and my bank balance. And they stored all the words of this book inside one of the wonders of our age—a personal computer.

///// Bugs and Bits

Let's get the bugs out first. In 1943, IBM completed the construction of the Mark I calculator and installed it at Harvard University. Two feet deep, eight feet high, and fifty feet long, it noisily carried out its calculations with shafts, pulleys, wheels, perforated paper tape, electronic vacuum tubes (this was before transistors), and relays—electric switches that opened and closed in response to magnetic forces. Malfunctions were frequent, but one particular malfunction has become memorialized in computer jargon. After a breakdown one day in September 1945, a search revealed that a moth that had somehow gotten inside had been crushed by a closing relay, thereby stilling both the Mark I (temporarily) and the moth (permanently). To this day, errors in computer programs and systems are called "bugs," and the identification and removal of such errors is called "debugging."

Now a bit of counting. The Mark I counted to ten as you and I usually do: 1, 2, 3, 4, 5, 6, 7, 8, 9, 10. It used the decimal system, which uses ten digits (including zero). Computers today use the *binary* system, in which there are only two *bi*nary dig*its*, or *bits*, 0 and 1. Their count to ten looks very different: 1, 10, 11, 100, 101, 110, 111, 1000, 1001, 1010. Each digit to the left represents a higher

THE ORIGINAL COMPUTER "BUG"

The carcass of the moth responsible for the terms *computer bug* and *debugging*, whose sad story is told in the text, has been preserved for posterity. It was recently forwarded from a naval museum in Virginia to the Smithsonian in Washington, D.C., where it will soon be put on display for the general public. The tale of its demise has been reported many times, but the sources I consulted disagreed on whether the computer involved was Harvard's Mark I or Mark II. I remained puzzled until I located *Voice of America Interviews with Eight American Women*, edited by C. Mompullen. In this book Grace Hopper, who was a Navy lieutenant at the time but who later advanced to admiral, solves the mystery of the moth's fatal accident: "Yes," she remembered, "that was back on Mark I. It was in 1945. We were building Mark II—and Mark II stopped." Apparently the scientists were testing the completed parts of Mark II by hooking them up to Mark I. Although the moth was actually in Mark I, it was the Mark II that stopped. Thought you'd like that important and vexing question settled.

power of 2. Thus 1111111 in binary is 127 (64 + 32 + 16 + 8 + 4 + 2 + 1), and 10000000 is 128. Representing 127 in binary required seven bits, but 128 required eight bits—a full *byte.*

A set of bits—a set of 0's and 1's—can represent not only numbers but also, through coding, letters and symbols from a typewriter keyboard and even various manipulations of numbers and symbols. Thus numerical calculations, word processing, and other forms of data processing can all be performed by the writing and reading of bits—by moving 0's and 1's around a computer's memory.

The binary system is convenient for the engineer who builds a computer memory, because each individual component that stores a digit need have only two variations, one corresponding to 0 and one corresponding to 1. Permanent magnets conveniently have a memory with two variations (S-N and N-S). Although they must compete with other binary memory systems, such as semiconductors (current on or off) and optical disks (presence or absence of a surface pit), magnetic memories were essential for the development of modern computers and remain vitally important today.

MIT's Whirlwind, called "one of the most innovative and influential computer projects in the history of computers," introduced the magnetic core memory in 1953. The core memory was an array of thousands of small doughnut-shaped magnets threaded with wires that could write and read data bits into the magnets with electric currents. Although the magnets were doughnut-shaped, they were not doughnut-sized. Smaller than Life Saver candies, smaller even than Cheerios, the cores were about 2 mm in diameter in the 1950s and later became as small as 0.3 mm. Magnetic core memories remained the most reliable and inexpensive computer memories for nearly twenty years, but they were eventually displaced by another product of modern materials technology—the integrated circuit.

One major goal of computer technology from the 1950s on was

miniaturization. For this reason computer memories have been judged by their *bit density*—how many bits could be stored in a given amount of space. The first successful minicomputer, introduced by Digital Equipment Corporation in 1963, was about the size of a refrigerator. Although it would be considered huge in today's world of laptops, it was much smaller than the giant mainframe computers that had preceded it, and it used magnetic cores. But the drive for miniaturization continued, and the components of core memories, tiny doughnut-shaped magnets woven together with an array of fine wires, were becoming harder and harder to reduce in size. Integrated circuits, with many transistors and other devices fabricated on a fine scale on the surface of silicon "chips," each year produced semiconductor memories with higher and higher bit densities. Semiconductor memories were also faster than core memories—bits could be accessed and changed faster—and this was a criterion of importance even for large computers. By the early 1970s, core memories were history. They played such a crucial role in the development of computers, however, that the symbol of Boston's renowned Computer Museum is a magnetic core memory. According to the museum's literature: "Core memories provided computers with the first random access, high speed, reliable storage. This allowed computers to meet their potential as tools for amplifying the abilities of human beings. Many inventions spurred the Information Revolution, but none with the same profound effect."

Although magnetic core memories are long gone, there is today a huge and rapidly growing market for magnetic disks and tapes for computer data storage, a market far larger than the demand for core memories in their heyday. In fact, the dollar value of magnetic recording devices produced in California's "Silicon Valley" exceeds the dollar value of semiconductor devices produced there. Given that most magnetic tapes and disks formerly employed iron oxide as the recording material, some have suggested that a more appropriate name for this area would be "Iron Oxide Valley." But, as we'll soon see, other magnetic materials are beginning to replace iron oxide. It may be safer to name the area "Magnet Valley."

///// A Future in Films

In today's computers, semiconductors provide the primary memory in which most of the actual computing takes place, but magnetic disks provide the secondary memory, which stores many more bits. The semiconductor memory, known as RAM (random-access memory), is faster and it has a higher bit density than the magnetic disk memory. But the disk has a lower cost per bit, and it has the decided advantage that it doesn't lose its memory when the power is off. Moving data bits around takes only nanoseconds in RAMs, versus milliseconds in magnetic disks. Whenever I ask my Macintosh to make changes of several kilobits in its secondary memory (by clicking on "save" after typing a few new pages, for example), or move data from its secondary memory to its primary memory (by opening a file), the total process can take a few seconds. A little clock face appears on my screen to encourage me to have patience with my relatively slow, but crucially important, magnetic memory.

Transferring data into and out of small computers is usually done with floppy disks, so called because the magnetic recording layer is on a thin sheet of flexible plastic, even though nowadays most are encased in a stiff container and can no longer flop. (The "hard disk" within the computer instead uses stiffer metal or glass sheets.) Large computers use magnetic tape to input and output data. Before the arrival of desktop computers, the public image of computers was based on large mainframes, which have a pair of large tape reels in view and which look much like giant tape recorders. Magnetic tapes have a lower cost per bit than magnetic disks, but they require a relatively long time to access data, as you know if you've ever rewound a cassette tape to replay a song. A disk drive system is constructed (Figure 9.3) so that the head can move to a particular point with much less motion than a tape requires, and so data can be accessed much more rapidly.

Magnetic recording on tapes and disks has been the dominant technology for electronic data storage for over forty years. Steady increases in bit density and decreases in cost per bit have made magnetic recording a rapidly moving target for competitive tech-

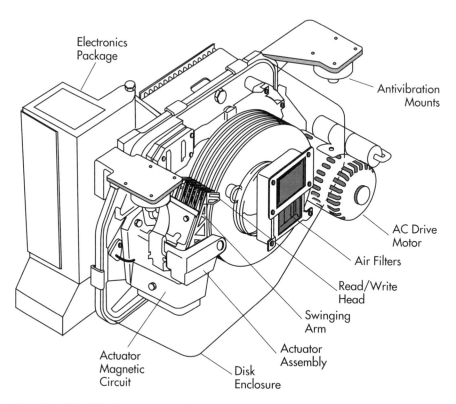

Figure 9.3 IBM "Winchester" 8-inch disk pack. To access data on the disks, the motor spins the disks and the actuators move the read/write heads across the disks. (An actuator, like a motor, converts electrical signals into motion, but only over a limited range.)

nologies. Commercial disks today can store over 50 megabits per square centimeter, which is more than 100,000 times the bit density in the first computer disk system, introduced by IBM in 1957. Bit densities over 100 megabits per square centimeter have been reported from the laboratory, and densities over 1 gigabit (one billion bits) per square centimeter are believed to be realistic in the near future.

Writing and reading data in magnetic disks at high bit density and high speed is no mean feat. High bit density requires the head to be extremely close to the disk. The relative speed between disks and heads today is about 100 miles per hour, and this is with the head flying about 50 nanometers above the disk. Experts in magnetic recording technology point out that this tiny separation requires remarkable control of the head position, analogous (if scaled up) to flying a jumbo jet a few millimeters off the ground! Luckily, most disks are flatter than the ground is (even to scale).

The drive to increase bit densities puts increasing demands on the permanent (hard) magnetic material in the disk, on the temporary (soft) magnetic material in the head, and on the overall design to minimize friction and avoid collisions of the low-flying head with the disk surface. Throughout the history of magnetic recording, fine particles of iron oxide have been the dominant choice of hard magnetic material. (This iron oxide, in recent years modified by the addition of a little cobalt, is mostly Fe_2O_3, rather than the Fe_3O_4 of magnetite, the main ingredient of lodestones. Since another form of Fe_2O_3 is the main component of rust, some disrespectful computer specialists refer to their magnetic disks, somewhat inaccurately, as rusty disks.) Today, however, the majority of computer hard disks instead use continuous metallic *thin films*, less than 100 nanometers thick, mostly of cobalt alloys. The soft magnetic materials of growing importance in recording heads are also thin films of magnetic metals. Metals have higher magnetizations than oxides, and thin films lend themselves naturally to high data densities. Most experts agree that the future of both disks and heads is in films.

Most large metallic parts for general engineering applications are made directly by casting liquid metal into a mold or by deforming or machining metal into final shape. It's not so easy to make metallic thin films that way. Thin films are usually made by depositing the desired atoms directly from a metallic vapor (produced by several different methods) onto a surface. Processing methods to deposit thin films of metals and other elements onto surfaces have become important in a wide variety of modern technologies, including magnetic recording.

Most heads are inductive heads: they use fact 8, the induction of electric fields by moving magnetic fields, to read data in magnetic disks and tapes. Others, however, use a different physical effect—a change in electrical resistance in the presence of a magnetic field, a phenomenon known as magnetoresistance (MR). An MR head can detect a bit of magnetic data as it passes by from this change in resistance, which for some materials can give a bigger electrical signal than induction. And there's no need for the head to have the complex coil geometry that inductive heads have. Some materials have recently been found that have a much larger magnetoresistive effect than previously studied materials, and there is much current interest in these "giant MR" materials, which consist of a very finely divided mix of two different metals. MR heads employing these new materials offer the promise of further increases in information storage density and may eventually replace inductive heads as read heads in many applications (but, since they lack the coil geometry of inductive heads, they cannot be used as write heads).

The major challenge to magnetic recording today comes from optical recording in the form of compact disks, in which bits of data are written and read by laser beam. Individual bits, 0 or 1, are represented by the presence or absence of a pit in the disk surface. Since the pit cannot be removed once formed, optical memories cannot be rewritten. They are read-only memories (CD-ROMs). Since they can be removed from the equipment used to record and read them, they are somewhat like floppy disks in their functions but have significantly higher bit densities. A hybrid technology called magneto-optic recording has arisen that combines the bit density available with lasers and the erasability available with magnetic recording. High-energy pulses of the laser cause magnetic reversal of small areas of the thin magnetic metal film on the magneto-optic disk, and a low-energy laser beam reads the presence or absence of such a reversed area. The future of this recording technology also appears to be in films.

One theme of this book has been the many different roles that magnets play in today's technology. Before leaving the topic of computers, I should therefore point out that, inside computers, magnets do a lot more than store the data in secondary memory. Magnets in the heads read and write the data. The motors that spin the hard and floppy disks, and the actuators that move the heads to very precise positions on the disks, are driven by forces between magnets. (Miniaturization and the proximity of these motors to the disks, especially in laptop computers, requires special materials, usually rare-earth magnets, and special designs to assure that magnetic fields from the motors will not damage data on the disks.) As in your television, magnets provide the forces that sweep the electron beams across the monitor screen to produce the image, and they provide the forces that generate sounds. Magnets, so essential to technology since the development of electrical power systems over a century ago, remain important in the most modern of technologies—the computer.

///// UP WITH MAGNETS!

10

///// Fighting Gravity with Levity

Does gravity get you down? Magnets can lift you up! Does friction slow you down? Magnets can speed you up! Fighting the forces of gravity and friction is one of the things that magnets do best.

Gravitational forces between objects are always attractive, but they are very weak unless the objects are massive. The gravitational force between two people is so small that it would require extremely sensitive equipment to measure it. (You may feel attracted to someone, but I assure you it's not gravity you feel.) But we live on the surface of a very large mass, and the resulting downward force of gravity—the attractive force between the earth and objects on its surface, like us, or the things we let go of, or perhaps Newton's apple—has a substantial and noticeable effect.

As useful as gravity is to us, one of the most popular fancies is to escape from it, or rather to be able to control the strength of its pull on us. Dreams of flying are common; I had them, and so did Carl Sagan, as he recounts in *Pale Blue Dot*, his recent book on space travel. Levitation is reported by mystics and demonstrated by magicians, and anti-gravity devices are extremely popular in science fiction. Fortunately, real anti-gravity devices are common. In fact, I'm sitting in one. The earth is pulling me down, but my chair is pushing me up with an equal and opposite force. If you pull the chair out from under me, the earth will temporarily have its way with me, but soon the floor will provide a balancing anti-gravitational force and I'll be stationary again.

Well, sure. Things I'm in contact with can push me up from underneath or pull me up from above. But can I oppose gravity with a force that doesn't require contact? At first look, hot-air and helium balloons seem to do it, and even 747s get off the ground. That's pretty neat, and sometimes even useful. But what if the object you want to levitate is like me—not lighter than air, and not winged and moving very fast through the air? Then you'll have to find another method. Fortunately, there is a way to provide anti-gravitational forces to lift things that are heavier than air, even if they're stationary, and to lift them even though they are not in contact with anything, including air. By now, you probably have guessed—you can do it with magnets.

And friction? Like gravity, friction has its uses. This is most obvious when it is nearly absent, as when my shoes or the wheels of my car are in contact with ice. Friction occurs whenever the surface of a moving object is in contact with the surface of a stationary object, and we rely on it to keep our bodies upright and our cars on the road. In other cases, however, friction wastes energy, heats up things you really don't want to heat, and wears away the surfaces of mechanical devices.

How can we combat friction? Lubrication is one approach. Another is to have objects rolling over each other rather than sliding—ball bearings are a very popular way to reduce friction. But suppose the moving object is supported by magnets and is not in contact with anything—just levitated. No contact, no friction. Not only can magnetic levitation fight gravity, it can also fight friction.

Unlike gravitational forces, magnetic forces can both repel and attract. North poles repel north poles and south poles repel south poles (fact 2). Several magnetic levitation (maglev) systems systems employ a vertical repulsion force between like poles to oppose the persistent downward pull of gravity.

A simple example often shown in children's science books is the repulsion between two ring-shaped permanent magnets mounted together on a rod or pencil. A more sophisticated example involving several magnets is a rotating toy such as "Revolution." In neither case, however, is the levitated magnet totally free of contact (see Figure 10.1). If you remove the pencil on

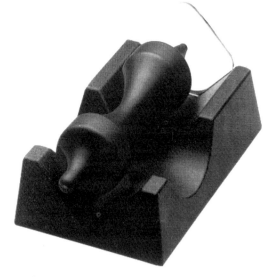

Figure 10.1 When two ring magnets are placed on a vertical support with like poles facing *(left)*, the upper magnet is levitated by the repulsive force between the two magnets. Repulsion is also at work in the maglev toy called "Rotation" *(right)*. Magnets in the base repel two ring magnets in the levitated rotating part, which is blocked from lateral motion by a glass plate. The levitated part can rotate for many minutes, because the only friction is between the glass plate and the steel-pointed end of the rotor.

which the ring magnets are mounted, the free magnet will flip over and opposite poles will mate. If you remove the glass plate in contact with one end of the rotating magnet in "Revolution," the free magnet will lurch forward and fall. Earnshaw's Theorem wins again.

About 150 years ago, an English scientist named Earnshaw stated that an object cannot be stably suspended in space by permanent magnets alone. No matter how hard you try, no matter how clever you are, and even though permanent magnets today are far stronger than they were in Earnshaw's day, it can't be done. People claim to have done it, and some even have been granted patents on their claims, but it can't be done. (Earnshaw's Theorem has since been proven mathematically, but it is not intuitively obvious, and people keep trying to get around it.) To keep the object suspended and stable, you need an added force—a force that was supplied by the pencil or by the glass plate in the examples shown in Figure 10.1. But didn't I say just a few paragraphs ago that magnets can support an object "not in contact with anything"? Yes, contactless levitation is possible—if you use AC electromagnets, superconductors, or a trick of modern technology known as active bearings. But Earnshaw was right—levitation is not possible with permanent magnets alone.

Fact 2 was enough to provide levitation in systems like two ring magnets held on a pencil. Other systems, however, provide levitation through fact 8, the induction of electric currents through changing magnetic fields. We call these *dynamic* systems, in contrast to the *static* systems of Figure 10.1.

I formerly had in my office a dynamic maglev demonstration that used an old phonograph turntable. The turntable was covered with a sheet of copper and a permanent magnet was mounted on the tone arm where the phonograph needle usually is. When the turntable was stationary, the magnet rested on the copper sheet. When the turntable was turning, however, the magnet would rise about a centimeter above the moving copper sheet. The relative motion between the magnet and the copper produced changing magnetic fields in the copper. Through fact 8, this induced electric currents in the copper, making it a temporary electromagnet that repelled the permanent magnet. Dynamic maglev resulting from the relative motion of a magnet and a conductor is a popular means of lifting high-speed trains.

You can also get dynamic maglev without moving your magnet or conductor, since you can get changing magnetic fields from an *AC* electromagnet. In Figure 10.2 you can see the amazement on the faces of two schoolchildren as they witness an aluminum pie pan mysteriously levitated above a large wooden box. Inside the box is a large AC electromagnet. The alternating currents produce alternating magnetic fields. The alternating fields induce currents in the aluminum, making the pie pan a temporary electromagnet, which is repelled by the electromagnet in the box.

The gentleman amazing the schoolkids in Figure 10.2 is Paul Thomas, who occasionally acts as "Mr. Magnet" to demonstrate the wonders of magnetism to local schoolchildren. In another dynamic maglev demonstration he performs, the relative positions of the AC electromagnet and the conductor are reversed: an electromagnet hovers above a slab of aluminum. The repulsion between an AC electromagnet and a conductor is mutual, although the effect is much more startling when the electromagnet is hidden and the levitated object is a common pie pan.

Yet another maglev demonstration using repulsion has recently become popular—the levitation of a permanent magnet above a high-temperature superconductor (or vice versa). For maglev to be achieved via currents induced in copper or aluminum, magnetic fields must be constantly changing, via relative motion or alternating currents. Otherwise electrical resistance would cause the induced currents to decay and disappear. But superconductors have no resistance, and currents induced by the approach of a magnet do not decay or disappear. They continue to flow on the surface of the superconductor, producing a field that repels the magnet. This is a static system like those in Figure 10.1, with no need for continually varying fields—though you do

Figure 10.2 Paul Thomas of MIT amazing two Cambridge schoolkids by magnetically levitating an aluminum pie pan. Within the box underneath the pan is a large AC electromagnet, which induces currents in the pie pan via fact 8 and creates an upward repulsive force on it.

have to keep the superconductor cold. Once it warms above its critical temperature, the currents will decay and levitation will cease.

Superconductors display very complex behavior in magnetic fields, complex enough to provide occasional surprises. A few years ago, physicists levitating small pieces of high-temperature superconductor above a rare-earth permanent magnet thought they'd lost a piece. Searching for it, they were surprised to find it suspended in midair *below* the magnet! This broke a well-known rule. For reasons I discuss below, stable suspension with attractive magnetic forces just doesn't happen—with ordinary magnetic materials. But superconductors aren't ordinary! The stable suspension of superconductors under magnets is now understood, but it wasn't predicted. And this exception to the rule allows me to emphasize the rule—stable maglev by attractive forces is not possible without special controls.

With the exception of the superconductor suspended under the magnet, all the maglev methods I've discussed until now, both static and dynamic, achieve levitation with *repulsive* forces. Achieving levitation with *attractive* forces is possible—and has been done recently to run Germany's maglev train, the Transrapid. But it's not as easy, because of the way magnetic forces vary with distance.

Whether repulsive or attractive, forces between magnets decrease with increasing distance. An object being levitated by a repulsive magnetic force moves up until the magnetic force pushing it up equals the gravitational force pulling it down. At smaller separations, the magnetic force is stronger than the gravitational force. At larger separations, it's weaker. Any departure from the equilibrium separation will lead to a net force up or down, and the object will return to the separation where the two forces are equal and opposite. Although the levitated object may not be stable against sideways or rotational motion (Earnshaw again), in repulsion it is stable against vertical motion.

Not so with attraction. When James Bond turned on the elec-

MAGLEV METHODS

- Repulsion between like poles of permanent magnets or electromagnets
- Repulsion between a magnet and a metallic conductor induced by relative motion
- Repulsion between a metallic conductor and an AC electromagnet
- Repulsion between a magnet and a superconductor
- *Attraction* between an electromagnet and a magnetic material, with sensors and active control of the current to the electromagnet used to maintain some distance between them

tromagnet in *The Spy Who Loved Me,* Jaws and his steel teeth were pulled upward. The closer he got to the magnet, the stronger was its pull. He ended up firmly attached to the magnet, not hanging in midair. If you're attracted to something, it's hard to keep your distance!

If Bond had wanted Jaws to hang in midair, he would have needed more control of the magnetic force. Once Jaws was a few feet off the ground, Bond could have decreased the current to the magnet until Jaws stopped rising and started to fall, then increased it again until he started to rise again, and continued cycling the current up and down to keep Jaws in midair. Attractive maglev works only if you maintain this kind of *active* control of the attractive force.

(Strictly speaking, the term *levitation* usually refers to pushing something up with repulsive forces. Pulling something up with attractive forces is *suspension*. So, strictly speaking, the German Transrapid train is not a maglev train but a "magsusp" train. That's just one example of why I don't always speak strictly.)

Fortunately, active control doesn't always require you to keep watching the suspended object and cycling the current by hand. Clever engineers have designed systems with sensors that measure the height of suspension and feed back electronic signals to control the current to the electromagnet. Systems like this, called "servo" or "feedback" systems, control the speed of automobiles (cruise control), the temperature in houses (thermostat), and many industrial processes. And they're important in the maglev devices we consider next—magnetic bearings.

///// Bearing Up

Maglev trains still seem rather exotic, and it will be many years before they are introduced into commercial service. But you already have a small maglev device in action at your house or apartment. It's your electric meter—more precisely, your watt-hour meter. Watts are a measure of electric power, the rate at which you use energy. Therefore watt-hours are a measure of the total electrical energy used, for which your friendly power company sends you a bill every month. If you're reading this book under a 100-watt bulb, and enjoying it so much that you keep reading it, and rereading it, for 10 hours (you just can't put it down), you will have used one kilowatt-hour, which will cost you about 7 cents in my town. Worth every penny!

Your watt-hour meter, like motors and generators and many other devices, has a rotating part and stationary parts. Reducing the friction between the rotating and stationary parts is the job of the bearing, which in the watt-hour meter is partly a magnetic bearing (Figures 10.3 and 10.4). The rotating disk is supported vertically by the repulsive force between two permanent magnets, much as the upper ring magnet was supported in Figure 10.1. That's not enough for complete stabilization, said Mr. Earnshaw, so guide pins at the top and bottom of the rotating shaft limit motion in the horizontal direction, much as the pencil kept things stable in Figure 10.1. Although levitation in the meter is

Figure 10.3 Watt-hour meter, in schematic cross-section. The rotating eddy-current disk (which measures the electrical energy consumed in your home) is levitated by the repulsion between like poles of the magnets below it. It is constrained from lateral motion by support pins at the top and bottom.

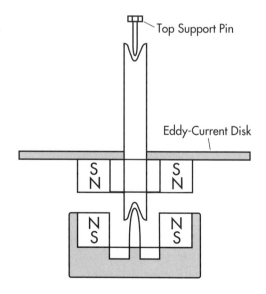

Top Support Pin

Eddy-Current Disk

Figure 10.4 The consumer's view of the watt-hour meter. The disk is projecting from the front plate, just below the label "KILOWATTHOURS."

therefore not completely contactless, the contact is very slight, and it results in very low friction and wear. That's important for a meter that is designed to last for many years with the disk in continuous rotation.

For completely contactless magnetic bearings, you must have active control of the magnetic forces in at least one direction. In Figure 10.5 an iron sphere is suspended below an electromagnet that is attracting it upward. A horizontal light beam continuously monitors the position of the sphere and controls the current to

Figure 10.5 Active magnetic suspension of an iron ball by *attraction* to an electromagnet above it. A light beam between the two side pieces detects the position of the ball and controls the current to the electromagnet to keep the ball suspended in mid-air.

the electromagnet to keep the sphere from rising or falling, as James Bond would have to have done to keep Jaws aloft. This is a very simplified version of the active magnetic bearings used in aerospace and industrial applications today. In most high-tech devices, though, several electromagnets are required to control motion in several different directions.

Because they are contactless and therefore virtually frictionless, magnetic bearings are especially useful for applications requiring very high speeds of rotation. One early application was in centrifuges, devices that use the centrifugal force created by spinning to separate materials of different mass density. In the 1950s, Jesse Beams of the University of Virginia pioneered the development of magnetic bearings in which small rotors could be spun at speeds of many millions of revolutions per minute. His ultracentrifuges have been used in the separation of isotopes, nuclei of the same element with different atomic weight.

Among other industrial applications of magnetic bearings are pumps for producing high vacuums, turbines for liquefying gases, and high-speed machine tools. Rotors as heavy as 10 tons, spinning thousands of revolutions per minute, have been supported by magnetic bearings—bearings that require no lubrication and produce no abrasion or wear. On a smaller scale, magnetic bearings, which have the advantage of very precise position control, have been used for patterning integrated circuits and in microscopes capable of atomic-scale resolution.

Satellites and space vehicles must be made of long-lasting, maintenance-free components that operate in a vacuum, an environment incompatible with lubricants (they'd evaporate) and in which metal-to-metal contact can lead to rapid wear and even adhesion. Magnetic bearings are a natural choice, and they have done service in supporting flywheels and gyroscopes used to guide and control the orientation or "attitude" of satellites. Magnetic bearings have also found related application in the inertial-guidance systems that control the trajectories of guided missiles.

Although magnetic bearings usually support cylindrical rotating parts, they have also been used to support flat platforms. In the Agfa Gaevert factory in Germany, a 6-ton table holding a film-winding machine is suspended only by invisible magnetic fields. It is called the "flying carpet" by factory workers. Germany now also appears likely to become the first country in the world to offer an intercity mass-transit "flying carpet"—in the form of a maglev train.

///// Flying Trains

As noted in Chapter 5, Bell Labs scientists reported in 1961 that a compound of niobium and tin maintained its superconductivity even when carrying large currents in the presence of large magnetic fields. That discovery changed superconductivity from an interesting scientific curiosity to a phenomenon with technological promise. Within a few years, superconducting high-field electromagnets were an engineering reality, and engineers began imagining applications for these new wonders. I recall attending several brainstorming sessions at General Electric at that time, and I'm sure that many other organizations held similar meetings. One idea, published in 1966 by Gordon Danby and James Powell of Brookhaven National Laboratory, was to use the high magnetic fields generated by superconducting electromagnets to produce magnetically levitated trains. An exciting idea that arose at the height of the anti-war movement of the '60s, "Mag Lev Not War" was a timely rallying cry in engineering circles.

In 1970, I was invited to talk about the materials science of high-field superconductors at a conference of Canadian physicists. To bring my paper to an exciting conclusion, I predicted that "in the 1980s we will probably be able to travel swiftly and comfortably between major cities on a superconducting train . . . supported by magnetic fields."

I was a bit over-optimistic. Full-scale maglev cars were built in the 1980s, and demonstration rides were available on short test tracks. But twenty-five years have passed since my talk, and intercity maglev trains are not yet a reality. The development of new transportation systems depends not only on technological progress but also on economic and political considerations, things that, in my naivety, I did not take sufficiently into account back in 1970.

Nearly all of the levitation methods described earlier in this chapter have been considered for mass transportation. Vehicles levitated by AC electromagnets and induced currents in a conductor (the principle in Figure 10.2) were studied by Emile Bachelet, a French engineer, as early as 1914. Although technically feasible, this approach was found to be impractical because it consumes huge amounts of electrical energy.

Levitation of significant weights via repulsive forces between two permanent magnets (Figure 10.1) was virtually impossible with magnetic steels and alnicos because of their low coercivities. Putting two like poles in close opposition tended to demagnetize one or both magnets. This approach became much more feasible in the 1950s, however, when the Philips Research Laboratories in The Netherlands developed hard ferrites. With their much higher coercivities, the ferrites were widely used in motors and speakers and held out the promise of being suitable for magnetic bearings and even maglev trains.

In his 1966 book *New Applications of Modern Magnets*, British engineer Geoffrey Polgreen vigorously promoted the use of hard ferrites for the construction of a magnetic railway he called Magnarail. From bricks of ferrite permanent magnets, he constructed a model track 12 feet long and a platform 28 inches long, each 18 inches wide. Levitated above the track, and stabilized laterally by siderails to satisfy Mr. Earnshaw, the platform could "carry the weight of a man" and be propelled along the track with modest forces. (Actually, the only photo I have seen of his model shows the platform carrying the weight of a young *woman*. But why quibble?)

From the data he gathered with his model, Polgreen estimated the costs of a full-scale Magnarail with 10-ton cars carrying 50 people (5 tons of people and 5 tons of magnets and support). He very optimistically concluded that track construction, vehicle construction, and vehicle operation would all be less expensive for the Magnarail than for a traditional railway. And speeds up to 300 miles per hour seemed feasible. The Westinghouse Electric Company did later build a small maglev model based on permanent magnets, and several other engineers have made optimistic calculations about such systems, but in recent years no one seems to have seriously considered this approach for full-scale maglev trains.

One problem not discussed by Polgreen is the likelihood that iron and steel objects would be attracted to the permanent-magnet tracks and cars, like a linear scrap heap. Considering the small gap between the track and a loaded car, and the brittleness of ferrites, one carelessly discarded screwdriver could cause considerable damage. The rare-earth magnets developed in the 1970s and 1980s have much higher coercivities and energy products than ferrite magnets and would make a permanent-magnet maglev system more technically feasible. Unfortunately, their high cost would make them less economically feasible. What's more, they're almost as brittle as ferrites and, being much more powerful magnets, much more likely to collect errant screwdrivers.

The maglev approach suggested by Danby and Powell, and adopted with modifications by the Japanese National Railways, is similar in principle to the phonograph model I described earlier. The model had a permanent magnet on the tone arm levitated above the copper sheet on the moving turntable. The Japanese have superconducting magnets on the cars and copper (or aluminum) coils in the guideway. Though different in detail, in both cases the relative motion of the magnets and the copper induce currents in the copper (fact 8 again) that produce magnetic fields to levitate the magnets. AC electrical power supplied to electromagnets in the guideway produces attractive and repulsive forces that propel the train, in a way fully analogous to the forces that drive the rotor in an ordinary electric motor (Chapter 7). The guideway thus serves as the stator of a linear motor. Magnets both levitate and propel maglev trains.

Test runs of vehicles using this system were started in 1977 on a four-mile test track on Kyushu, Japan's southernmost island. In December 1979, the unmanned ten-ton ML-500 achieved a world speed record (for trains) of 321 miles per hour. Tests continued with a revised track design and the MLU001, a prototype of a commercial maglev car that was tested at high speeds in two-car and three-car units. The MLU002, with 44 seats, carried thousands of passengers on low-speed demonstration rides before it was destroyed by fire in 1991. (An hour after the blaze, a firefighter approaching the car reportedly had his fire axe pulled from his hand by a superconducting magnet that apparently had somehow escaped the heat.) The MLU002 was replaced by the fireproof MLU002N (Figure 10.6, *top*), which in February 1994

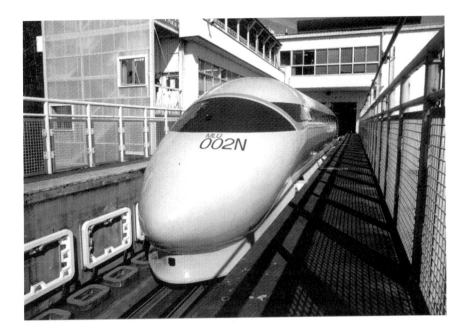

Figure 10.6 Two maglev train systems that have been operated at full scale. The Japanese MLU002 *(top)* contains superconducting magnets and is levitated by repulsive forces produced by currents induced in the guideway coils by the moving train. Germany's Transrapid *(bottom)* is levitated by attractive forces between copper-wound electromagnets and the steel track, the principle demonstrated in Figure 10.5.

reached 268 mph, running unmanned. Since dynamic maglev depends on velocity, the MLU002N starts on wheels and does not "take off" until accelerated to about 100 mph. At maximum speed it levitates four inches above the track, but it remains at maximum speed for only a few seconds before electromagnetic braking begins. A four-mile track doesn't allow much cruising time for a high-speed maglev vehicle.

Experimental runs continue on Kyushu, but progress toward a possible commercial maglev line in Japan now depends on the construction of a much more ambitious test track to be built on the main island of Honshu. It is planned for a mountainous region about halfway between Tokyo and Nagoya. Most of its 22 miles will be in tunnels, and the track will include curves, grades, and viaducts. Part will be double-tracked, to allow testing of the effects of two high-speed trains passing each other. If construction of this track and subsequent test runs are successful, these 22 miles will eventually become part of a maglev line linking Tokyo and Osaka, providing a speedier alternative to the already speedy "bullet train" now linking those two cities. But if this line is eventually built, it probably won't become the world's first commercial maglev line. That will probably be built in Germany.

Germany's Transrapid (Figure 10.6, *bottom*), like the iron sphere in Figure 10.5, operates by servo-controlled attraction between an electromagnet and a temporary magnet—magsusp, not maglev. The bottom of the cars wraps around the edges of a T-shaped track and is attracted upward by magnetic forces to a controlled gap of about three-eighths of an inch. As with Japan's MLs, magnetic forces also provide propulsion. Based on a design conceived in the 1920s, today's Transrapid 07 is the latest prototype in a program started in 1969.

Testing has been carried out since 1983 on a track near the German-Dutch border. The track has a five-mile straightaway and turn-around loops at each end, with a total track length of nearly 20 miles. Many thousands of passengers have taken demonstration runs around the track. When the Transrapid 07 reached 271 mph here in December 1989, it was a world record for a maglev train carrying passengers.

Encouraged by many years of successful tests, the German government recently decided to construct a Transrapid maglev line linking Hamburg with Berlin. This choice has strong political symbolism, since Berlin not so long ago was an isolated outpost deep within East Germany. The estimated cost is nearly 9 billion marks (about 6 billion dollars), two-thirds of which will be provided by the government. Trains will reach a maximum speed of 250 mph between stations and will cover the full 180-mile route in less than an hour. Much faster than traveling by plane, if you count the time at each end required to get to and from the airport. The line is expected to attract many more passengers than the 10 million per year needed to make it profitable. Track construction is scheduled to start in 1996, and the target for operation is 2005. Maybe the Transrapid will meet this schedule, maybe not, but it looks today like the best bet to become the world's first intercity maglev.

Other maglev systems using attractive magnetic forces include the Japanese HSST, demonstrated in a low-speed version at Expo '86 in Vancouver, and a commercial system at Birmingham, England, linking the airport and train station. The Birmingham system has been in continual operation since 1984, but its top speed is a ho-hum 26 mph.

What about the United States? The most ambitious maglev program here in the 1970s was the Magneplane, designed at MIT by Henry Kolm and Richard Thornton. A 1/25 scale model was built and tested, but funding dried up in 1975. Hopes were revived in 1991, when Congress authorized a $725 million maglev program as part of the $151 billion Intermodal Surface Transportation Efficiency Act (known as "IceTea" inside the Beltway). The four groups awarded grants all had proposals employing superconducting magnets. Teams headed by Magneplane International, Bechtel, and Foster-Miller proposed electrodynamic (fact 8) systems based on repulsive forces, while the team led by

Grumman proposed a system based, like the Transrapid, on electromagnetic attraction.

But a funny thing happened on the way through Congress—the appropriation committee refused to fund the program. Maglev proponents blame Congressman Bob Carr, a Democrat from Michigan who apparently believes that high-speed trains are a threat to the auto industry. Prospects for maglev may improve soon, since Carr left the committee in 1995.

Until federal funding returns, the future of maglev in the United States is in the hands of private groups and state legislatures. Supporters have formed the High Speed Rail/Maglev Association, an advocacy coalition that promotes all high-speed rail alternatives, including both maglev and wheel-on-track systems (like Japan's "bullet train" and the French TGV). Maglev studies are underway in Florida, New York, Pennsylvania, and several other states, but it is not yet clear where or when the first U.S. maglev line will be built.

Joseph Vranich, president of the High Speed Rail/Maglev Association, convincingly presented the case for high-speed trains in his highly praised 1991 book *Supertrains: Solutions to America's Transportation Gridlock.* To quote from the book's preface, "What if someone were to invent a magical new mode of transportation that was safe, energy efficient, and environmentally benign? Someone has. It's called the train."

I've enjoyed traveling on high-speed trains in Europe and Japan. I look forward to the possibility of traveling on a maglev "flying carpet" between Berlin and Hamburg and perhaps between Tokyo and Osaka. But I live in the United States, the world's scientific and technological leader. Is it too much to hope for a maglev train between Boston and New York?

///// MAGNETS AT WAR

11

///// Hitler's Secret Weapon

In September 1939, when Hitler set off World War II by invading Poland, he bragged that his "secret weapons" would win the war for Germany. Two months later, off the coast of England, mysterious explosions proved that this was no idle threat. In three days, hundreds of lives were lost as a dozen ships blew up in shallow waters—waters in which no submarines had been detected and which had been swept for the normal variety of contact mines (mines moored to remain just below the surface and explode on contact with a ship's hull). Hitler's secret weapon had claimed its first victims; his "Mine Kampf" had begun.

A few nights later, English plane spotters saw German aircraft circling over the broad mouth of the Thames River and dropping cylindrical objects that floated down on parachutes and disappeared into the water. The next morning, a fisherman reported something unusual resting on a sandbank off a nearby town. By lucky accident, one of Hitler's deadly secret weapons was in full view, and the mine experts of the British Admiralty soon arrived to investigate.

The object on the sandbank was 8 feet long and 2 feet in

diameter, but it had no mooring chains, like those attached to contact mines, and no propeller or steering gear, like those on torpedoes. One brave expert, suspecting a magnetic trigger, removed all keys and magnetic objects from his pockets and approached the mysterious device with nonmagnetic tools. His associates stayed at a safe distance, and he described to them each step he took, so that, if he were blown up with the mine, others would know what not to do on the next occasion. Fortunately for him and for Great Britain, he successfully dismantled the device. For this 1,200-pound mine containing about 650 pounds of high explosives, the crucial piece of the complex triggering mechanism was a small magnet much like a compass needle. With the mechanism automatically activated after the mine settled to the bottom, the needle would respond to any large steel object, like a ship, that came within about 40 feet of it. Movement of the magnetic needle would close an electric circuit that detonated the explosive. The mine was designed to explode directly under a ship, precisely where the ship was most vulnerable.

Once they understood how the bombs were triggered, the British scientists set about developing countermeasures. Ships were girdled with current-carrying cables, and the electric currents were adjusted to cancel any magnetic fields extending from the ship. This process, called "degaussing," made the ships relatively immune to magnetic mines, but it was expensive and time-consuming and the electric currents required frequent adjustments.

Methods were also developed to explode magnetic mines before they could damage any ships. Tugboats, carefully degaussed, towed long cables carrying electric currents provided by powerful generators. The currents flowing through the cables created magnetic fields (fact 6) in the water behind the tugboats and triggered the mines harmlessly. Other ships were outfitted with huge electromagnets that created magnetic fields far ahead of the bow, exploding the mines safely before the ship passed over them. However, the tugboats with the "electric tails" were found to be the most effective approach.

By late 1940, countermeasures were sufficiently developed that German magnetic mines were no longer a major threat. Many naval experts believe that one of Hitler's most vital mistakes of the war was to employ this secret weapon too soon and too sparingly. Had the magnetic bombs been held back for a while, and then released in far greater numbers along with heavy aerial bombing in preparation for the planned invasion of Britain, they would probably have been much more effective. Their limited use early in the war allowed the Allies sufficient time to develop countermeasures that rendered Hitler's first secret weapon relatively ineffective.

Although the United States did not officially enter the war until the December 1941 bombing of Pearl Harbor, American magnetics experts were in England in 1940 helping the British scientists combat the magnetic mines. One of those sent over was Francis Bitter of MIT, whose work on powerful electromagnets was discussed in Chapter 4. Later Bitter returned to Washington to design magnetic mines and other "influence mines" (some triggered by sound, some by water pressure, some by a combination of effects) for Allied use against the Japanese. Mines were a major factor in the success of our naval war in the Pacific; each new variation was able to sink many ships before countermeasures could be developed.

Another magnetic weapons program that Bitter worked on during the war was the design of a target-seeking torpedo. A magnetic detector within the torpedo automatically directed the torpedo toward a ship once it came within range of the ship's magnetic field. This became a forerunner of the many target-seeking missiles that followed. Bitter also contributed to devices for magnetic detection of submarines from the air—the magnetic-anomaly detectors (MADs) that, after the war, helped American companies prospect for minerals and oil and helped geologists demonstrate sea-floor spreading and continental drift.

///// Hunting for Red October

The first successful use of a submarine in naval combat was in 1864, when a Confederate sub blew up a Union ship in Charleston harbor. The following decades saw intensive submarine development by the navies of many nations, particularly Germany. The submarine had become a major military weapon by World War I, in which German U-boats (*Unterseeboote*) sank over 10 million tons of Allied and neutral shipping. Much more advanced submarines prowled the seas in World War II, but they faced much more sophisticated techniques of anti-submarine warfare, including detection by MADs.

One of the most famous phrases identified with the war years is the terse "Sighted sub, sank same." The hard part was the sighting. Bitter and others considered many techniques for locating vessels beneath the waves, and since submarines, like the surface ships sunk by Hitler's secret weapon, are large steel objects that produce small local changes in magnetic field, magnetic detection was naturally one of these techniques.

The U.S. Navy first put MADs into active service in early 1944. Catalina aircraft carrying MAD gear, appropriately called "Madcats," located and sank two U-boats in the Straits of Gibraltar shortly after being put into action. In Navy parlance, the detectors were magnetic airborne detectors rather than magnetic-anomaly detectors, but they were still MAD.

Operational details of items like anti-submarine equipment do not often appear in the unclassified literature. Assuming Tom Clancy's description of MAD gear in *The Hunt for Red October* is correct, however, airborne MADs can detect a submarine as far as 600 yards on each side of the airplane's path, but no farther. Thus, these MADs are probably most useful for patrolling limited stretches of water, like the Straits of Gibraltar, or when some other technique, such as sonar or radar, has established the general location of an enemy submarine.

With the development of missile-carrying nuclear submarines capable of remaining submerged for months at a time, sonar became a vital strategic tool not only for "shooting wars" but for Cold War rivalries as well. Active sonar locates underwater objects by radiating sound waves and detecting their reflection from the objects—listening for the echo. In early sonar systems, sound waves were generated and detected by electromagnets with cores of nickel, a temporary magnet that changed length when magnetized—a phenomenon known as *magnetostriction*. Today most sonar systems instead use piezoelectrics—materials that change length in response to electrical fields rather than magnetic fields. But magnetic materials far more effective in generating sound than nickel (compounds of iron and rare-earth metals), and others far more effective in detecting sound than nickel (amorphous metals), have recently been developed. An experimental sonar device generating sound with an iron and rare-earth compound, driven by an electromagnet wound with a high-temperature superconductor, has recently been constructed by American Superconductor Corporation. Perhaps, before long, magnetic materials will once again play an important part in sonar. Peaceful uses of sonar include depth determination, location of schools of fish by fishing fleets, and locations of shipwrecks (with, for the lucky, lost treasure).

A colleague of mine told me about an unusual use of permanent magnets to detect submarines during the Cuban missile crisis of 1962, when the United States had established a naval blockade around Cuba to interrupt the Soviet Union's supply lines to the island. Alnico magnets were imbedded in plastic rings about two feet in diameter, and aluminum fins or "flappers" were attached to the rings. When American naval personnel thought a Russian sub might be lying on the bottom, silent, they would drop a number of these assemblies overboard. The rings would attach to the steel subs and the ocean currents would move the flappers, generating sound that could be detected by sonar.

Back in World War II, submarines had to surface every twenty-four hours to charge their batteries and get fresh air for their crew. Once surfaced, they could be detected from miles away by airplanes carrying airborne radar, developed late in the war. Airborne radar was made possible by a British "secret weapon," the *cavity magnetron.*

///// Magnetrons and Radar

Much as sonar detects objects by their reflection of sound waves, radar detects objects by their reflection of radio waves—by "radio echoes." The birth of radar is usually dated to January 1935, when the British Air Ministry, alarmed by the build-up of Hitler's air force, formed the Committee for the Scientific Survey of Air Defence. This group was directed to consider all possible means of combating enemy bombers, including directed energy beams, popularly known as "death rays."

Robert Watson-Watt, head of the Radio Research Laboratory, was asked by the Committee whether there was any possibility of concentrating enough energy in a radio beam to incapacitate a pilot or his aircraft. He reported back, on the basis of calculations by his assistant Arnold Wilkins, that a radio death ray was impossible but that *detection* of enemy aircraft by reflected radio waves looked quite feasible. Although Watson-Watt is often called the "father of radar," he notes in his autobiography that paternity is always more questionable than maternity, and calls Wilkins the "mother of radar."

By 1938, the "baby" of proud parents Watson-Watt and Wilkins was operational, and a chain of radar stations along Britain's eastern coast could detect aircraft up to a distance of 100 miles.

(The locations of the radar towers, by government decree, "should not gravely interfere with the grouse shooting." The British, many still doubting that war would come, had their priorities.) After the fall of France in June 1940, the German Air Force began heavy air attacks on England, sometimes sending out more than a thousand aircraft a day. But early radar detection of approaching planes enabled the outnumbered Royal Air Force to win the "Battle of Britain" and destroy Hitler's plans for the invasion of England.

This early radar system emitted radio waves at a frequency of about 30 megahertz (30 million cycles per second), corresponding to a wavelength of about 10 meters. The British soon realized that the accuracy and utility of radar could be greatly increased if waves of higher frequency—with wavelengths on the order of centimeters rather than meters—could be used. Workers at Birmingham University were assigned the task of developing a radio transmitter capable of "many watts on few centimeters." The result was the cavity magnetron, a tiny device that could be held in the palm of a hand (Figure 11.1) but was capable of delivering thousands of watts at a frequency of 3 gigahertz (3 *billion* cycles per second) and a wavelength of 10 centimeters. Such short-wavelength radio waves became known as *microwaves* (see Figure 8.2).

In the summer of 1940, a delegation of British scientists visited the officially "neutral" United States. Demonstration of their cavity magnetron, which generated far more microwave power than any American devices, made a great impression, and the Radiation Laboratory was formed at MIT for the specific purpose of developing microwave radar.

In the cavity magnetron, the magnetic field from the permanent magnet and fact 9 are used to get an electron beam traveling in a circular path (as in Figure 5.1B) past the openings of several cavities in a block of metal. The movement of the electron beam generates high-frequency electromagnetic oscillations in the cavi-

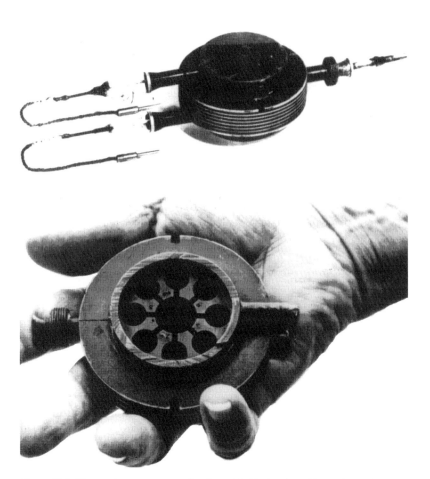

Figure 11.1 The cavity magnetron, developed in England and brought to the United States in 1940. Called "the most valuable cargo ever brought to our shores," it generated high levels of microwave power. and made possible the development of radar systems of great importance to the Allies in World War II. The assembled magnetron *(top)* was put between the poles of a permanent magnet (not shown); an interior view *(bottom)* shows the configuration of the cavities.

ties, somewhat analogous to the sound vibrations set up in a Coke bottle when you blow across its mouth. In the early days of the MIT Radiation Laboratory, a group of physicists gathered around a table that held a disassembled magnetron, and their group leader, I. I. Rabi, told them, "It's simple. It's just a kind of whistle." Another prominent physicist, Edward U. Condon, responded, "Okay, Rabi, how does a whistle work?" Condon reports that Rabi could not come up that day with a satisfactory explanation, but the high-powered group at the Radiation Laboratory eventually understood the magnetron and microwave radar systems well enough to have a major impact on modern warfare.

From intensive work at MIT and elsewhere, microwave radar became an extremely effective weapon for both defensive and offensive purposes; it was used for directing gunfire, for air and naval navigation, for discriminating one aircraft from another, and for detecting ships from the air. Airborne radar detected submarines from long distances even if just the periscope was above the water. It became "the one single weapon that defeated the submarine and the Third Reich," according to a very reliable authority—Grossadmiral Karl Doenitz, commander of Hitler's U-boat fleet. The cavity magnetron, brought to America by British scientists in 1940, was called by one American expert "the most valuable cargo ever brought to our shores."

The Germans first became aware of the newest secret weapon in February 1943, when they downed a British plane carrying microwave radar. They immediately set up a laboratory to develop magnetrons and microwave radar, but their success came too late to affect the outcome of the war. German progress was severely hampered by shortages of cobalt and nickel, important ingredients in the alnico permanent magnets that provided the field for the magnetrons produced by the Allies. Gerhard Hennig, a magnet expert who worked in the German navy during World War II, calls these "the magnets that decided the war." (He had earlier encountered British alnico magnets when the Germans were investigating the 1943 disabling of their battleship *Tirpitz*. A British minisub, undetected, had entered the Norwegian fjord in which the *Tirpitz* was anchored and attached explosives to the underside of the battleship's hull with alnico magnets.)

Magnetrons, the secret weapon that gave the Allies radar superiority over Germany in the closing years of the war, now power microwave ovens. Radar, of course, also has many peacetime uses, including air-traffic control and weather surveillance. The least popular application is probably the police radar gun, which caught even Watson-Watt, the "father of radar," one day in 1954 as he was hastening to a speaking engagement. The incident generated considerable news coverage and the following anonymous verse:

> Pity Sir Robert Watson-Watt
> Strange target of his radar plot
> And thus, with others I could mention,
> A victim of his own invention.
> His magical all-seeing eye
> Enabled cloud-bound planes to fly
> But now, by some ironic twist,
> It spots the speeding motorist
> And bites, no doubt with legal wit,
> The hand that once created it.

At first, alnico permanent magnets provided the magnetic field for magnetrons. (The magnetrons in today's microwave ovens mostly use hard ferrite magnets.) Later, magnets with higher coercivities were used to guide electron beams in "traveling-wave tubes," devices that serve as amplifiers in microwave radar systems. The first major use of high-coercivity rare-earth permanent magnets, introduced commercially by G.E. in the early 1970s, was in traveling-wave tubes used in air combat in Vietnam. At the time, I was both working on magnets at G.E. and opposing the Vietnam War by participating in anti-war demonstrations. I was therefore pleased that the new magnets rapidly found a peaceful application—in electronic wristwatches.

///// Calutrons and Little Boy

It was 8:15 A.M. Hiroshima time, August 6, 1945. High above the city, an American B-29 bearing the name *Enola Gay* unloaded a four-ton bomb, about 10 feet long and 29 inches in diameter, known to workers in the Manhattan Project as Little Boy. Inside the falling bomb, four radar units bounced signals off the approaching ground. When the radio echoes indicated that the bomb had fallen to a height of 1,900 feet, the units sent a firing signal to the cordite explosive. The expanding cordite forced a "bullet" of uranium-235 into a "target" of the same metal, and Little Boy, the world's first combat atomic bomb, exploded. The death and destruction in Hiroshima, which still horrifies the world but surely hastened the end of World War II, was caused by about 100 pounds of uranium-235 produced in Oak Ridge, Tennessee, in an array of 900 electromagnets.

(In the summer of 1995, the *Enola Gay* and a model of Little Boy were put on display at the National Air and Space Museum in Washington. Plans for a more comprehensive exhibit questioning the right and wrong of dropping the bomb were scrapped after heated protest, especially from veterans' groups. Fifty years have passed, but emotions remain strong on this issue.)

The nuclear fission of uranium had been discovered by German physicists shortly before the war. By 1942, American scientists were in a desperate race to create an atomic bomb before Hitler's Germany did. The electromagnetic method of isotope separation was one of several approaches they explored. Physicists in both Allied and Axis countries knew that a neutron hitting a U-235 nucleus would cause fission of the nucleus and that the products of that fission would include, in addition to lots of energy, at least two neutrons. The production of more neutrons

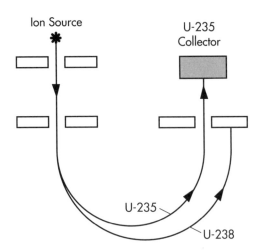

Figure 11.2 How the calutron separated fissionable U-235 isotope from the predominant U-238 isotope. A beam of uranium ions, aligned by passing through two slits, entered a tank (not shown) with a magnetic field perpendicular to the plane of the drawing. As in Figure 5.1B, the ions traveled in circular paths (fact 9), but the radius of the path of the lighter isotope was slightly smaller than the radius of the U-238 path. The U-235 isotopes were intercepted in their path and trapped in the collector.

with each fission could create a runaway chain reaction, leading to tremendous destructive power. All that was needed was perhaps 50 to 100 pounds of uranium consisting primarily of U-235. The catch? Most of the atoms in naturally occurring uranium are the heavier isotope, U-238. Less than 1 percent are U-235. (Isotopes are atoms of the same element with different numbers of neutrons in the nucleus. All uranium nuclei contain 92 protons, but U-235 nuclei contain 143 neutrons while U-238 nuclei contain 146.)

Ernest Lawrence, winner of the 1939 Nobel Prize for Physics for inventing the cyclotron, championed the idea of using electromagnets to separate U-235 from U-238. Uranium ions, like electrons and other charged particles, travel in circles when they are released perpendicular to a constant magnetic field (fact 9). The greater the mass of the ion, the greater the radius of the circular path it follows. U-238 ions, because their nuclei have more neutrons (and are therefore heavier) than the nuclei of U-235 ions, would follow one path when in the presence of a magnetic field while the U-235 ions would follow a different path (Figure 11.2). Lawrence first demonstrated the separation of a minuscule amount of U-235 in a 37-inch-diameter cyclotron at the University of California in Berkeley. He then scaled up the experiment with a 184-inch cyclotron and its 4,500-ton electromagnet, at that time the largest in the world. Results were promising enough that a massive plant for electromagnetic separation was then built at Oak Ridge, Tennessee. The devices were named *calutrons* after their origin at the University of California. (The suffix *-tron*, a particular favorite of physicists, has been attached to dozens of electronic devices both large and small.)

Lawrence's 900 calutrons consisted of large vacuum tanks in which the uranium ions moved, surrounded by iron-core electromagnets providing the magnetic field. With copper in short supply because of war demands, the windings of the electromagnets were mostly of silver—over 10,000 tons of silver worth over 300 million dollars! The calutron factory, housed in buildings covering more area than 20 football fields, in fall 1944 started turning out enriched U-235 at a rate of a few pounds per week. (When you separate things atom by atom, material accumulates slowly!) By summer 1945, enough had been accumulated to power Little Boy—a bit too late for it to be used against Germany, but soon enough to encourage the surrender of Japan.

An alternate technique, gaseous diffusion, also contributed to the accumulation of U-235 at Oak Ridge, and plutonium, produced from U-238 in nuclear reactors, was the fissionable material that powered the Trinity test bomb in New Mexico and the bomb dropped on Nagasaki. But most of the U-235 that destroyed Hiroshima, and awakened the world to the atomic age, was separated by the magnets of Lawrence's calutrons. Magnetic separation continues today at Oak Ridge, yielding isotopes for medical and other peacetime applications.

From Hitler's mines in 1939 to America's atomic bomb in 1945, magnets were behind many of the new and deadly weapons of World War II. (A multitude of magnets also played their traditional, but less sensational, roles in motors, actuators, meters, speakers, and other devices; each B-29 bomber, for example, employed 420 permanent magnets.) The outstanding success of microwave radar, generated by magnetrons with alnico magnets, prefigured the high-tech electronic warfare of today. From ballistic missiles to "smart" bombs, from Stealth aircraft to Star Wars, magnets are essential components of modern weaponry. I hope that this aspect of the magic of magnets will become obsolete someday, but that day may be far in the future.

///// MAGNETS AT PLAY

12

///// Child's Play

"Have fun with magnets! Amusing! Educational! Scientific! Mystifying!" That's how one toy catalogue describes its magnetic offerings. I agree—they're all of the above! Although this book is not designed for children (as I said in the preface), many magnets are. Toys and games based on magnets are so common that even many pre-schoolers have learned a few "facts about the force."

Many toys and games are based simply on the attractive force between a permanent magnet and iron (fact 3). One toy my children enjoyed many years ago had a cartoon of a man's face and some iron filings under a transparent plastic cover. With a "magic" magnetic wand, they could move around the iron filings and give the man a moustache, a beard, eyebrows, and a variety of hair styles. This toy is still in stores today, as are more sophisticated variations. In the newer versions, with names like Ghostwriter or Magnadoodle, fine magnetic particles are suspended in viscous liquids and magnetic "pencils" of various shapes can be used to draw pictures or make patterns.

One toy store owner told me that her most popular magnetic

item remains the letters of the alphabet mounted on small permanent magnets. (At least the letters are popular with parents, who usually make the purchasing decisions.) Adult versions are marketed in which the magnets are mounted with words instead of letters, and in sets with different themes, including "romance" and "hot romance," the latter explicit enough that your kitchen may become X-rated. A classier version, named Magnetic Poetry, containing 440 words, can be found at many museum gift shops. This "toy" has been declared "a fabulous way of seeing how words rub against each other" by Jonathan Galassi, president of the Academy of American Poets.

Many "travel games"—magnetic checkers, magnetic chess, and so on—have small permanent magnets in the pieces and some iron in the board. Attractive magnetic forces hold the pieces in place throughout bumpy car, plane, or train rides. I also recently received a set of magnetic playing cards, which were promoted as "windproof." In this case, the board is a weak permanent magnet and the cards, containing iron, are temporary magnets. (Permanent magnets in the cards would have made them hard to shuffle!) The attractive force between the board and the cards also allows you to play solitaire in bed even when the board is vertical.

A magnet-based game that my wife and daughters like, I think because they usually beat me, is a fishing game with a rotating "ocean" that contains eight plastic "fish." The fish contain small steel rods, and each player attempts to catch as many fish as possible with a tiny fishing pole and a magnet that dangles from the end of the string. This is not as easy as it sounds, because the mouths of the fish are continually opening and closing, and good timing is essential.

A number of interesting toys are based on fact 5—that iron objects in contact with a permanent magnet become temporary magnets themselves, and can then attract other iron objects. (This is the principle behind paper-clip chains, mentioned in

Figure 12.1 "Magnetic sculpture" toy. A strong permanent magnet in the base magnetizes the iron figures, which then attract each other and so may be positioned in a variety of gravity-defying arrangements.

Chapter 1.) These toys have numerous small iron pieces that do not mutually attract until they come under the influence of a large and strong permanent magnet in a plastic base. The individual pieces may be shaped like hearts, dollar signs, dinosaurs, acrobats, aerobic dancers, you name it. When in contact with the base, the attractive forces between the many temporary magnets allow the creative construction of "magnetic sculptures" (Figure 12.1).

Toys with two or more permanent magnets can display the effects of both the attractive forces between unlike magnetic poles and the repulsive forces between like poles (fact 2). One popular shape for these permanent magnets is a sphere, exemplified by "the original Geospace magnetic marbles." Even a simple set of brightly colored magnetic marbles can provide hours of challenging fun to a curious mind, but you can also purchase Geostax, a "magnetic marble building set," and Geozooms, "magnetic-powered critters." With these, you insert the magnetic marbles in plastic mounts shaped like beams and bricks or like assorted "critters," which will then attract or repel depending on the orientation of the magnetic poles.

Also fun are sets of O-shaped disk magnets, usually magnetized so that the poles are on the disk faces. These will perform a number of simple but entertaining tricks, including an impressive display of magnetic levitation (Figure 10.1). Some rare-earth magnets are so strong that one hidden in the palm can produce surprising attractive or repulsive forces on another one placed on the back of the hand, an impressive demonstration of fact 4 (magnetic forces can act at a distance through nonmagnetic barriers).

Attractive and repulsive forces between permanent magnets have also been used to devise three-dimensional puzzles. A catalogue description of one says, "You'll be attracted to this colorful wooden puzzle, but make sure the magnetic pieces don't repel your efforts to assemble it." Repelled by this description, I didn't order it.

Magnets are a popular topic of books and kits designed to introduce science to young children. Consider *Explorabook*, "A Kids' Science Museum in a Book" developed by the Exploratorium of San Francisco. Although it covers many areas of science, it starts with 15 pages on magnets, covering most of our "facts about the force" with discussion of various experiments and demonstrations. It even includes a "dinnertime magic" trick that involves attaching a strong permanent magnet to your knee and pressing it against the underside of the dinner table. As we'll see in the next section, some Russian "psychics" have apparently already learned this trick.

One children's science kit I recently purchased was an electric motor kit "for age 8 and up." Although I am very much in the "up" category, I must confess it took me several hours of painstaking effort to wind the rotor electromagnet wire around the iron core, assemble the motor, and convince it to operate. I wonder how many 8-year-olds were discouraged from a career in science and technology by this devilish kit.

Fact 8, the induction of an electric field by changing magnetic fields, is important to a number of "perpetual motion" toys, such as the "Rolling Dolphins" toy shown in Figure 12.2. Started at one end of the ramp, the wheel of dolphins will roll back and forth, with no sign of tiring, for months. Friction should slow it down, but it "just keeps going." What's the secret? A battery hidden in the base serves as the energy source, and permanent magnets in the dolphins and a coil of wire in the center of the base do the rest. Each of the three dolphins contains a permanent magnet. As the wheel rolls past the center of

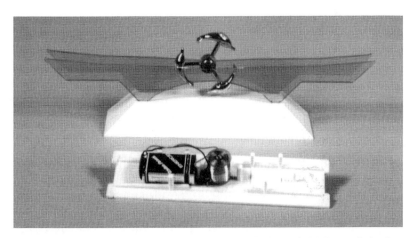

Figure 12.2 "Rolling dolphins" toy. The three-armed wheel of dolphins rolls back and forth for months without slowing. The foreground shows what is hidden in the base: the coil of an electromagnet and a battery to provide the energy needed to overcome friction.

the ramp, the moving magnets produce a changing magnetic field in the coil of wire hidden in the base. This induces a voltage (fact 8) that triggers a transistor circuit, which allows the battery to deliver a larger *reverse* current to the coil. The magnetic field produced by this current (fact 6) produces a force on the permanent magnet that keeps the wheel moving. Ingenious.

In another device, a similar mechanism keeps a tiny top wandering and spinning endlessly (it seems) on a shallow plastic base. The tiny top contains an even tinier permanent magnet, and the base, like the base for the Rolling Dolphins, contains a coil of wire, a battery, and a transistor circuit. This device, called "Top Secret," can be thought of as an electric motor, with the top as the rotor. The transistor circuit is designed so that the coil voltage induced by the spinning top (fact 8 again) results in a large reverse current from the battery. This current produces a magnetic field (fact 6) and a force to keep the top spinning.

On my most recent visit to the Museum of Science in Boston, the windows of its museum store featured Top Secret, Rolling Dolphins, and several related toys—all in motion with no visible sources of power. They all appeared to be violating one of the basic tenets of science—the impossibility of a perpetual-motion machine. "Well," you'll say, "they all have hidden magnets." True, but magnets alone can't produce a perpetual-motion machine, as many inventors over the centuries have discovered, to their dismay. What is needed to provide continual motion is a source of energy, which in these toys is provided by the hidden battery. These toys will all run down when their batteries run down, but that takes many weeks—a span of time that seems almost "perpetual" in today's world of short attention spans.

In *De Magnete*, Gilbert ridiculed various authors who had claimed to describe perpetual-motion machines based on magnets. His closing words were harsh: "May the gods damn all such sham, pilfered, distorted works, which do but muddle the minds of students!" Nearly four centuries have passed since Gilbert published this diatribe, but patent offices are still besieged by countless inventors of ingenious magnetic devices purported to produce more energy than they consume. Hope springs eternal.

///// Magnets in Fiction

The attractive and repulsive forces between magnets have inspired not only toymakers and the engineers who have designed our technology but also many writers, whose imaginations need not be bounded by established experimental facts. An early example is Laputa, the "island in the air" that appears in Jonathan Swift's classic satire, *Gulliver's Travels* (1726). Isaac Asimov called it "true science fiction, perhaps the earliest example we have of it."

Laputa was levitated by means of a giant lodestone 6 yards long and 3 yards thick. Apparently interacting with magnetic material underneath on the continent of Balnibarbi, one side of the stone was attractive, the other repulsive. By tilting the stone, the island could be moved sideways, downwards, or upwards, to a maximum height of 4 miles. Although, as Asimov stated, "Swift's mechanism to keep Laputa aloft and in motion wouldn't really work," the use of repulsive magnetic forces to achieve levitation does not seem quite so fanciful in these days of maglev trains.

The idea of using magnetic forces to counteract the pervasive downward force of gravity appears much earlier than Swift's tale, in Pliny's *Natural History*. Pliny reports that the architect Timochares had designed for Ptolemy II (285–246 B.C.) a temple with a lodestone vault so that an iron statue contained within it would be suspended in midair. The temple was to honor the pharaoh's wife Arsinoe, who was also his sister. (Both Ptolemy and Arsinoe proudly took the name Philadelphus, for "brotherly love.") Pliny reports that both Timochares and Ptolemy died before the temple was completed, avoiding what might have become overwhelming engineering difficulties. Stories also circulated that Mohammad's tomb in Mecca was similarly magnetically levitated, but Gilbert in *De Magnete* derided the idea as mere fable.

A much more recent example of maglevity appeared in the comic strip "Dick Tracy." In the late 1960s, creator Chester Gould had Tracy and his pals flying around town in one-man maglev vehicles. (These anti-gravity devices, like Swift's Laputa, could reach heights of miles, not just the mere inches reached by the highest of today's levitated trains.) Gould was so enraptured with the concept of maglev in particular, and the technological potential of magnetism in general, that his comics frequently bore the dramatic statement: "The nation that controls magnetism controls the universe."

Although Jonathan Swift and Chester Gould appeared to be most fascinated by the repulsive forces of magnets, other writers focused on their *attractive* forces. In one of the thousand-and-one *Tales of the Arabian Nights,* a ship carrying the Prince is blown off course by heavy winds, and the Captain tells him:

> Tomorrow by the end of day we shall come to a mountain of black stone, the Magnet Mountain, for thither the currents carry us willy-nilly. As soon as we are under its lea, the ship's sides will open and every nail in plank will fly out and cleave fast to the mountain; for that Almighty Allah hath gifted the lodestone with a mysterious virtue and a love for iron, by reason whereof all which is iron travelleth towards it; and on this mountain is much iron, how much none knoweth save the Most High, from the many vessels which have been lost there since the days of yore.

It happened as the Captain predicted, and many were drowned in the seas in front of the Magnet Mountain. Fortunately, the Prince survived to tell the tale (via Scheherazade).

Attractive magnetic forces were also important in a much more recent adventure story—the movie *Star Trek VI: The Undiscovered*

Country. A Klingon ship carrying emissaries to a peace conference was hit by torpedoes that disabled its artificial gravity system. As unprepared Klingons floated about, two men were beamed aboard from the Enterprise, wearing magnetic boots. Whereas magnetic forces opposed gravity in *Gulliver's Travels* and "Dick Tracy," here they instead provided an effective substitute for gravity, since the floors of the Klingon ship were apparently good temporary magnets. The attraction between the boots and floor allowed the two men to walk about freely, shooting the floating and relatively helpless Klingons.

Later in the film, a search of the *Enterprise* uncovered the magnetic boots and, eventually, the villains who were trying to sabotage the peace conference. In the scene in which the boots were discovered, their magnetic nature was demonstrated by their adherence to a cabin wall. That suggests, at least at first glance, that the walls of the *Enterprise,* as well as the floors of the Klingon ship, were made of steel, like the sides of my refrigerator. That bothered me a bit. Steel is used sparingly in today's airplanes and space vehicles because it is so much heavier than aluminum, titanium, and various composite materials. But I probably shouldn't worry about such things. Even though they'll be working with the same hundred-odd elements that we have now, the engineers of the future will have learned to put them together in different ways to produce very different properties. Besides, *Star Trek* is only fiction. Or so some people say.

One world of fiction seldom mistaken for reality is the world of superheroes and supervillains that appears in the pages of Marvel Comics. The X-Men, superheroes whose exploits also appear each week on television, have, as their oldest and most dangerous enemy, Magneto—the "master of magnetism." Operating from Asteroid M, a colossal space station orbiting earth, Magneto uses his mastery of magnetism, which he draws from "the primal forces of the earth itself," to pursue his goal of dominating hapless mankind.

I have throughout this book referred to the "natural magic" of magnets, but magnets are also responsible for some "unnatural magic" as well. It has long been realized that magnetic forces can enhance the art of illusion. Jean-Eugene Robert-Houdin (1805–1871) was a popular French magician now considered to be the father of modern conjuring. (The American magician and escape artist Houdini borrowed many tricks, as well as part of his name, from this master.) One of Robert-Houdin's most famous tricks is known today as the "light and heavy chest."

He would announce to his audience that he could "deprive even the most powerful man of his strength." Eventually someone muscular would accept his challenge. The magician would hand him a wooden box, and it would be obvious that neither man had any difficulty lifting it. Robert-Houdin would then set the box on the floor and, with mysterious hand motions and incantations, would cast his spell. "Now you are weak," he would say, and to the amazement of all, the man could no longer lift, or even budge, the wooden box! An electromagnet hidden under the floor had been turned on, attracting an iron plate concealed within the box with enough force to thwart even the strongest man. Robert-Houdin could, of course, restore the man's strength at any time by turning off the current to the magnet. It helped that audiences of his time were totally unfamiliar with electromagnets.

I recently purchased a small magic trick that multiplies your money faster than any mutual fund. You simply place a small piece of red plastic over a penny and say the magic words. When you remove the plastic, the penny has miraculously changed into a dime! The "penny" looks just like a real penny, but is actually a hollow bit of steel which fits easily over a dime. A permanent magnet is hidden in the plastic, so that when you remove the plastic you also remove (and quickly palm) the "penny," revealing the dime.

I also acquired a plastic "King Tut" who normally rests peacefully in a plastic sarcophagus. Once he is removed from the coffin, however, the pieces are in the power of the "magician" and mysterious forces make it impossible to return him to his resting place. A permanent magnet, of course, has been embedded in each piece, and the position of the one in the coffin can be shifted. By making the right magic gestures, you can turn the mobile magnet in the coffin so that the pole that repels or attracts the King, as you wish, is in position to do your bidding.

You've probably seen a magician seemingly cut a rope in two and then magically restore it to its original uncut length. Most such rope tricks depend on sleight of hand and misleading knots, but rare-earth magnets now allow another option. Tiny "neo" magnets in the ends of two halves of a rope, attracting each other with a force of several pounds, can make the rope appear continuous. The magician can then "cut" the rope between the two magnets (with non-magnetic scissors!) and later magically rejoin the two halves, undoing the effects of the "cut."

Many other magic tricks, big and small, rely on hidden magnets. (See, for example, *Magnetic Magic* by Paul Doherty and John Cassidy.) But the owner of my local magic store refused to sell me any more, since the ethics of the magic business require that you don't explain how the tricks work. I assured him that the readership of my book would not be large enough to ruin the futures of David Copperfield, Doug Henning, et al., but he remained firm—no more tricks.

Some illusionists do not admit to being illusionists—they instead claim to have supernatural psychic powers. A prominent example in the 1970s was Uri Geller, a handsome Israeli stage performer who made quite a splash in the United States. One of his many "psychic" powers was psychokinesis, the ability to move objects with his mind. One of the objects he was able to move most impressively was a compass needle. Martin Gardner, a noted science writer who enjoyed debunking charlatans, saw Geller perform this feat on Tom Snyder's television show in 1975. Gardner noticed that, although Geller made many distracting hand motions, the needle deflected whenever Geller's head was near the compass. Gardner concluded that a small permanent magnet was concealed "either in his mouth or sewn in his shirt collar." Other popular locations for hidden magnets that he identified included the tip of a shoe and under the trousers on the knee, where they can be easily manipulated when one's legs are hidden under a table on which a compass—or other movable iron object—is located. Gardner added: "But I'm already in trouble with some of my magic friends for revealing too much."

Shortly after I read Gardner's piece on Geller, I happened to see a prime-time television show on Russian psychics. Now that the Iron Curtain was down, the announcer exclaimed, we would finally be able to see these gifted Russians demonstrate their amazing powers. Several were specialists in psychokinesis. Having read Gardner, I was able in most cases to guess where their magnets were hidden. If those Russian psychics had gotten close enough to the Iron Curtain, they probably could have moved *it*, too.

Compass needles and iron and steel objects (see Figure 12.3) can be moved by hidden permanent magnets more easily since the development of rare-earth magnets in the 1970s. Even a magnet small enough to conceal under your fingernail can move a compass needle from a considerable distance. Improvements have made magnets more magical than ever, but the magic remains natural magic—not supernatural.

The natural magic of even rare-earth magnets, however, is probably not as powerful as claimed by some companies marketing magnetic devices today. A company in Pennsylvania sells "Magnalawn 2000, a magnetohydrodynamic fertilizer utilizing

ROBOTMAN by Jim Meddick

Figure 12.3 As this recent syndicated comic strip implies, a psychic may demonstrate his "telekinetic" powers by means of magnets hidden under shirt cuffs or trousers. (Reprinted by permission of NEA, Inc.)

chemical-free natural magnetic energy." Containing several permanent magnets, this remarkable device attached to your garden hose supposedly will "neutralize the harmful chemicals found in water, reduce water consumption up to 40%, and promote healthier soil and growth conditions." Other magnetic products claim to soften tap water and reduce corrosion of water pipes. It is true that magnetic forces could change the distribution of impurity ions (charged atoms or molecules) in water and have some subtle effects on complex phenomena like corrosion. But I think I'll save my money until I see some scientific proof that such devices actually work.

Even more likely to be bogus are magnetic devices that are claimed to boost the fuel efficiency of your car (while simultaneously eliminating toxic tail pipe emissions) or sharpen razor blades. Another product is said to improve the flavor of wine. It is a doughnut-shaped magnet through which you pour the wine. Pouring the wine over the north pole produces a "rich and tangy" taste, and pouring it over the south pole yields a "smooth and sweet" taste. The price is forty-five dollars, which wine experts say is about forty-five dollars too much. Caveat emptor.

Professional illusionists like Uri Geller demonstrate psychokinesis not only by moving compass needles but also by bending spoons and other objects, supposedly using only the powers of their minds. Martin Gardner, magician James Randi, and others have demonstrated how these tricks are done, but many prefer to believe that Geller and other spoon-benders have supernatural powers. On a recent TV interview program, Connie Chung featured real-life "Ghostbusters" who visited a house to rid it of poltergeists thought to be responsible for the mysterious displacement of various household objects. Detecting no poltergeists, the psychics concluded that the objects had been moved by psychokinesis. The psychokinesis had apparently been unintentionally produced by residents of the house, probably as a result of mental stress. Stress-reduction therapy was recommended. Connie Chung seemed impressed that this diagnosis had been provided free of charge. Psychokinesis is alive and well on your television screen.

Belief in psychokinesis, telepathy, clairvoyance, and other forms of extrasensory perception (ESP) is so widespread that some scientists have attempted to provide explanations for these mysterious phenomena. Since invisible electric and magnetic fields can produce forces, and even deliver television images through the air, they are commonly suspected of being the mechanism behind ESP. In a 1975 book entitled *Superminds,* John Taylor considered various possible explanations of the spoon-bending of Uri Geller, and other manifestations of ESP, and decided that electromagnetic waves were involved. ESP, he and many others believe, is a form of biological radio communication. Taylor, a professor of mathematics at the University of London, suggested that "the human operator emits the electromagnetic intentionality field which interacts with the receiver of a telepathic message or with a piece of metal." For other miraculous phenomena, such as ghosts, poltergeists, mediumship, and psychic healing, he proposed a similar explanation. (As discussed in Chapter 5, the strong fields from superconducting magnets can indeed sometimes produce "poltergeist" effects!) To be fair to Professor Taylor, I should note that he changed his mind about ESP several years later and concluded in a 1979 article in *Nature* that "the existence of any of the psychic phenomena we have considered is very doubtful."

Another questionable phenomenon often explained in terms of electromagnetism is dowsing, the supposed ability of some people to detect underground water with the use of wooden sticks, metal rods, pendulums, or other devices generally termed "doodlebugs." The annual conference of the American Society of Dowsers now draws a gathering of close to 1,000 men and women, many of them convinced that they have this power. Taylor and others have suggested that dowsers, like the MADs used to detect magnetic anomalies and thereby locate iron ores, oil fields, and submarines, are extremely sensitive to small variations in the earth's magnetic field.

I don't want to be accused of rejecting dowsing, ESP, and other "paranormal" phenomena as completely impossible, since our understanding of the human mind is still very limited. And, as we'll see in the Chapter 15, brain activity does produce minuscule but measurable magnetic fields. Until now, however, demonstrations of dowsing, psychokinesis, and other forms of ESP have failed whenever examined under carefully controlled experimental conditions—conditions that include the presence of an expert magician capable of detecting trickery. Many people will still believe in ESP, astrology, UFOs, communications with the dead, and other unproven phenomena simply because they want to. For me, the proven phenomena of nature—including the natural magic of magnets—are already wonderful enough.

///// MESMERISM AND MAGNETIC THERAPY

13

///// **Healing with Magnets**

At his birth in Switzerland in 1493, he was named Philippus Aureolus Theophrastus Bombast von Hohenheim. By his death in Salzburg fifty years later, he was known throughout Europe, much more simply, as Paracelsus. He chose the name to indicate his debt to Celsus, the first-century Roman physician and author of *De Medicina,* our major source of information on early Greek and Roman medicine. (*De Medicina* was published in 1478, one of the first medical works to be published after the invention of the printing press. Fortunately, it no longer takes fourteen hundred years to get a book published.)

Paracelsus was an alchemist as well as a physician and is generally credited with establishing the importance of chemistry in medicine. He is also usually credited (or discredited) with being one of the first to argue the miraculous healing powers of magnets. Magnets, after all, have the mysterious power to attract iron; perhaps, he reasoned, they can also attract diseases from the body! Paracelsus described detailed procedures, employing magnets, designed to transplant diseases from the body into the earth. His procedures were often successful, but their success probably depended less on the miraculous powers of magnets than on the amazing powers of the human imagination.

Paracelsus was not unaware of the important role of the patient's mind in the process of healing. He wrote, "The spirit is the master, the imagination is the instrument, the body is the plastic material. The moral atmosphere surrounding the patient can have a strong influence on the course of his disease. It is not the curse or the blessing that works, but the idea. The imagination produces the effect."

Although written over four hundred years ago, his ideas sound pretty modern. They could easily be quotes from Bill Moyers's 1993 book and TV series entitled *Healing and the Mind.* As psychiatrist Carl Jung (1875–1961) wrote, "We see in Paracelsus not only a pioneer in the domains of chemical medicine but also in those of an empirical psychological healing science."

Paracelsus was very controversial in his own time, but many physicians and healers of the ensuing centuries were strongly influenced by his writings. In particular, the development of strong carbon-steel permanent magnets in eighteenth-century England brought renewed interest in his ideas concerning the healing powers of magnets. Among those interested was Father Maximilian Hell, a Jesuit priest and Professor of Astronomy at the University of Vienna.

Father Hell tried treating patients with steel magnets made into different shapes corresponding to the body part that required healing. He claimed many successes, including curing the stomach cramps of a prominent baroness. In 1774, a Viennese physician and friend of Hell's used some of these magnets to treat a young woman suffering from a severe nervous illness. The friend was Franz Anton Mesmer, and Mesmer's success with the "magnets from Hell" led directly to the widespread promotion of his theory of "animal magnetism." His efforts soon made him one the best-known and most controversial figures of eighteenth-century Europe.

Mozart's popular comic opera *Cosi fan tutte* (Women are like that) premiered in Vienna in January 1790. Near the end of Act One, two young men pretend to take poison as part of a complex plot to test the loyalty of their fiancées. A maid to the young ladies then sings:

> Let's run and get a doctor.
> I know of one who's marvelous
> With people who are ill.
> He's known for working miracles
> Without a knife or pill.
> He's famous for his skill
> Perhaps he'll save them still!

She returns disguised as a doctor and produces a giant magnet from under her costume. She touches the foreheads of the two imaginary invalids with the magnet, then strokes the whole length of their bodies. Her accompanying song has been translated as: "Here and there a touch of the magnet, the stone of Mesmer, who was born in Germany and became so famous in France." The invalids rapidly recover, "thanks to the doctor's amazing technique."

Mozart's gentle spoof of his good friend Mesmer seems a bit obscure to today's audiences, but the scene required no explanation in 1790. In the preceding decade, much of Europe had been "mesmerized" by the controversies surrounding Mesmer's "animal magnetism." Mesmer's approach to healing had drawn supporters in prominent circles, among them the marquis de Lafayette and Madame Du Barry (mistress of King Louis XV), as well as influential critics.

Mesmer himself traced his theory of animal magnetism to his experience in 1774 treating the young woman mentioned earlier with steel magnets obtained from Father Hell. He reported that "when my patient had another attack I fixed two magnets of horse-shoe type to her feet and a heart-shaped magnet to her breast. Suddenly she felt a burning sensation spreading from her feet through all her joints like a glowing coal . . . and likewise from both sides of the breast to the crown of the head . . . the pains gradually went away, she became insensitive to the magnets. The symptoms disappeared and she recovered from the seizure."

Mesmer experimented with applying the magnets to different parts of the patient's body. She gradually became much improved, and Mesmer was convinced that the cure had been produced by his control of the ebb and flow—an "artificial tide"—of the "universal fluid" within her body. In his memoirs he explained that

> certain properties analogous to those of the magnet reveal themselves, especially in the human body. It is possible to distinguish different and opposite poles that may be changed, linked, destroyed or reinforced . . . This property of the human body, which makes it responsive to the influence of the heavenly bodies, and to the reciprocal action of the bodies around it, made me, in view of the analogy with the magnet, call it *animal magnetism.*

Although Mesmer at first used actual steel magnets, an approach he later termed "mineral magnetism," later experiments convinced him that the magnets were only serving as a "conductor" for the "universal fluid." He found he could "magnetize" paper, wood, leather, water—virtually anything—and produce the same effect on the patient. He concluded that the animal magnetism resided in himself, the various materials simply aiding the flow of the "universal fluid" between him and the patient.

A series of remarkable cures began to spread his reputation as a healer, and in 1775 and 1776 he toured Austria, Bavaria, Switzerland, and Hungary with notable success. Then as now, the prospect of being cured of afflictions "without a knife or pill" was very attractive. One patient who had gone deaf recovered his hearing when Mesmer held his hands over the patient's ears.

Others lost their chest pains or stomach spasms when Mesmer simply stroked the affected parts with his hands. His growing popularity made it difficult to provide prolonged individual attention to the many who wanted his care, and he experimented with group therapy using "magnetized" objects that several patients could touch at the same time.

Despite Mesmer's growing success, his demands for approval of his unorthodox theories were not well received by the conservative medical establishment of Vienna. This lack of acceptance and other problems convinced him to leave Vienna in 1778 and travel to Paris, where his fame had preceded him. He was an immediate sensation.

This period in Europe has been termed the Enlightenment—the Age of Reason. Mesmer's theories may seem very unenlightened and unreasonable today, but they took Paris by storm. Soon rich and poor alike thronged to his clinic and his famed *baquets,* wooden tubs containing "magnetized" water and iron filings. From each *baquet* protruded iron rods, which the patients grasped and touched to the afflicted parts of their bodies. To enhance circulation of Mesmer's magnetic "universal fluid," they linked fingers. Assistant "magnetizers" were on hand, but Mesmer himself would appear now and then with great flair. With dramatic gestures of his hands or a long iron rod, he would perform his cures, which often induced convulsive fits, a Mesmerian "crisis" that hastened healing.

Among Mesmer's many followers, as noted above, was the marquis de Lafayette, hero of the American Revolution. In 1784 Lafayette wrote to his friend and comrade in arms, George Washington: "A German doctor named Mesmer, having made the greatest discovery upon animal magnetism, has gathered students among whom your humble servant is called one of the most enthusiastic . . . Before leaving, I will obtain permission to reveal to you Mesmer's secret, which, you may believe, is a great philosophical discovery."

Mesmer, obviously hoping that the Lafayette-Washington connection would successfully introduce animal magnetism to the United States, followed with his own letter to Washington, which concluded: "It appeared to us that the man who merited most of his fellow men should be interested in the fate of every revolution that has for its object the good of humanity. I am, with the admiration and respect that your virtues have ever inspired me with, Sir, Your Obedient Servant, Mesmer." Washington's reply was polite but noncommittal. He was apparently unconvinced that America needed another "revolution."

Mesmerism was a great success with the French public but, as in Vienna, could not earn the respect of the official medical or scientific organizations. Orthodox physicians were threatened by Mesmer's success, and in 1784 King Louis XVI was persuaded to establish a Royal Commission to evaluate the claims of animal magnetism. Commission members included Antoine Lavoisier, the renowned chemist, Dr. Joseph Guillotin, whose name became associated with the device that beheaded Lavoisier ten years later, and Benjamin Franklin. Franklin's experiments with electricity had helped spark public interest in scientific developments, perhaps increasing the public's susceptibility to new ideas, such as Mesmer's, that were presented in scientific terms.

In a 1989 article in *Natural History,* Stephen Jay Gould called the Commission's report on animal magnetism "a key document in the history of human reason." The report opened very logically: "Animal magnetism might well exist without being useful, but it cannot be useful if it doesn't exist." The Commission soon learned that Mesmer's "universal fluid" had no detectable properties and could therefore be studied only through its effects. The investigators tried unsuccessfully to magnetize themselves and then turned to controlled experiments on a series of patients. Their results clearly showed that all the effects that were observed could be attributed to the power of suggestion, and that "the practice of magnetization is the art of increasing the imagi-

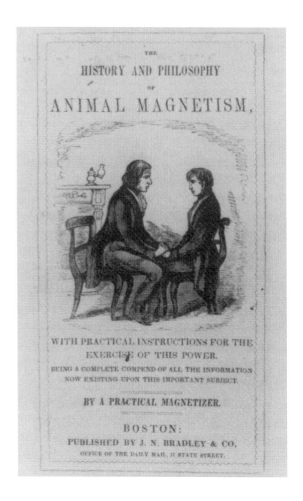

Figure 13.1 A book on animal magnetism published in Boston in 1843.

nation by degrees." Thomas Jefferson, arriving in Paris soon after the Commission report, noted in his journal: "Animal magnetism is dead, ridiculed."

Ridiculed, perhaps, but not dead. Mesmer himself faded from public view but many Europeans remained convinced of the powers of animal magnetism, and Mesmer's ideas did eventually cross the Atlantic. In Boston a "practical magnetizer" published a book on the subject (Figure 13.1), and Phineas Quinby of Portland, Maine, who had taken to Mesmerism in 1837, employed touch and suggestion in his magnetic healing practice. One of Quinby's patients cured by magnetism was Mary Patterson, who became, after her last marriage, Mary Baker Eddy, the founder of Christian Science. Originally a supporter of magnetic healing, she later decided that prayer was the sole source of healing and reviled "malicious animal magnetism," which she called "the action of error in all its forms." Another American who became interested in magnetic healing was Daniel David Palmer. Focusing on the laying on of hands, he opened Palmer's School of Magnetic Cure in Davenport, Iowa, in the 1890s. His ideas developed into the system of hands-on therapy known as chiropractic.

In Europe, several groups of magnetizers gradually evolved into students of hypnosis. The trance induced in many of Mesmer's patients is thought to be what is today called a hypnotic trance, and most dictionaries today consider mesmerism to be a synonym for hypnotism. Vincent Buranelli, one of Mesmer's biographers, writes that "scientific hypnosis would not have developed when it did except for him." Buranelli also concludes:

> Mesmer's tragedy was that he had the right facts and the wrong theory. If he had not had the theory of animal magnetism, he might have realized that his cures were psychological, and he might have carried the Western mind forward into psychosomatic medicine . . . His impress is clear on psychiatry, psychosomatic medicine, personality studies, and group therapy. He was the Columbus of modern psychology, which is enough of an achievement for any explorer.

The analogy between Mesmer and Columbus is a good one. They both were guided into a strange new world with the use of magnets, but both were unaware of where they had actually landed.

Lest you think that animal magnetism is only ancient history, I refer you to a 1987 book by Heinz Schiegl entitled *Healing Magnetism: The Transference of Vital Force.* Schiegl uses the laying on of hands and magnetic stroking—but no actual magnets. He does, however, recommend purchasing a compass, because it is "important to line the patient up along the magnetic meridian"—that is, north-south. The book's many photographs show not only how to apply magnetic healing to various parts of the body but also how to magnetize a glass of water, and even a bathtub full of water, with hand gestures. He warns that "for magnetic baths to be effective, it is important that the bathtub is situated, if possible, in the magnetic meridian, i.e., in a north-south direction. Deviations of a maximum of 15 degrees to the west or east are acceptable, but with larger deviations the effect is no longer guaranteed."

Although Schiegl appears to attribute more importance to the earth's magnetic field than Mesmer did, his book otherwise appears to be a detailed modern manual of animal magnetism. Jefferson was wrong.

///// Mineral Magnetism

In his book, Schiegl differentiates clearly between "magnetic healing," which uses only the hands, and "magnetic therapy," which uses actual magnets to produce effects on the body. Mesmer termed the latter "mineral magnetism."

In *De Magnete* (1600), Gilbert discussed various procedures claimed to employ the "medicinal virtue" of lodestones. Most involved grinding natural magnets into powder and either ingesting the powder or applying it externally in a salve or plaster. Gilbert ridiculed such claims: "Thus do pretenders to science vainly and preposterously seek for remedies, ignorant of the true causes of things." However, he clearly distinguished between lodestone as a magnet, which "attracts when it is whole, not when reduced to powder," and lodestone as a carrier of iron, which "is beneficial in many diseases of the human system." We know today that the human body contains about 5 grams of iron, mostly in the form of hemoglobin, which plays the crucial role of transporting oxygen from the lungs throughout the body.

The first man-made permanent magnets stronger than lodestones were developed in eighteenth-century England. Gowen Knight (1713–1772) achieved an international reputation, and a substantial income, by constructing powerful permanent magnets from steel. Knowledge of Knight's magnets reached Vienna and Father Hell, leading directly to Mesmer's animal magnetism. Knowledge of the fame and fortune achieved by Mesmer reached England and Dr. James Graham.

In Chapter 2, I noted various connections between magnets and romance. No one carried this concept further than Dr. Graham, who opened his "Temple of Health and of Hymen" in London's Pall Mall in 1781. (Hymen was the Greek god of marriage.) Graham's Temple featured several miraculous healing devices, including an "electrical bath" and a "magnetic throne." These could cure a wide variety of health problems, but patients who wanted help with their romantic life could also pay 50 pounds, later raised to 500 pounds (a huge fee in those days), for a night in his prime attraction—the famous Celestial Bed.

This remarkable bed was twelve feet long by nine feet wide and was supported by forty glass pillars, within which musical instruments provided celestial sounds. The mattress was filled with oriental spices, and the silk and satin sheets were "suited to the complexion of the lady who is to repose there." Advertised as an infallible cure for sterility, the bed derived most of its powers from magnets. Graham wrote:

> The chief elastic principle of my celestial bed, is produced by artificial lodestones. About fifteen hundred pounds weight of artificial and compound magnets, are so disposed and arranged, as to be continually pouring forth in an ever-flowing circle, inconceivable

and irresistably powerful tides of the magnetic effluvium, which every philosophical gentleman knows, has a very strong affinity with the electrical fire. These magnets too, being pressed give that charming springyness—that sweet undulating, tittulating, vibratory, soul-dissolving, marrow-melting motion; which on certain critical and important occasions, is at once so necessary and so pleasing.

"Loving stones" indeed!

Graham further assured potential customers that under the "invigorating influences of music and magnets . . . strong, beautiful, nay double-distilled children . . . must infallibly be begotten." With that sort of guarantee, perhaps a fee of 500 pounds was not unreasonable; children begotten under ordinary circumstances are so unpredictable!

Visits to the Temple and its Celestial Bed were often enlivened by erotic performances by "half-naked wenches." One of Dr. Graham's attractive magnetic performers was young Emma Lyons, who later became Lady Hamilton and achieved renown as the mistress of Lord Nelson, naval hero of the Battle of Trafalgar.

The nineteenth century brought the development of the electrical industry and a host of electromagnetic wonders. The telegraph, telephone, phonograph, and electric light gave convincing evidence to the general public of the mysterious powers of electric and magnetic fields, and America's Midwest became especially fertile ground for "doctors" touting magnetic therapy. The most notable of these was Dr. C. J. Thacher, whom *Collier's Magazine* dubbed "King of the magnetic quacks."

According to an 1886 mail-order catalogue distributed by Thacher's Chicago Magnetic Company, magnetic therapy provided a "plain road to health without the use of medicine." Iron in the blood was the magnetic conductor of the body, and its action could best be enhanced by wearing magnetic garments. The company provided a full line of such garments: a complete costume contained over 700 magnets, which provided "full and complete protection of all the vital organs of the body."

Thacher himself wore a magnetic cap, magnetic waistcoat, magnetic stocking liners, and magnetic insoles. He claimed that magnets could cure anything: "Let the authorities turn over ten cases to me. I'll put my magnetic shield on 'em and restore the harmonious actions of the brain, and everything will be well! Paralysis? An easy problem. Had five cases . . . Cured 'em right off. Winked. Spoke. Got up and walked. Paralysis? Pish!"

Moving closer to the present day, consider Sue Wallace, a "Doctor of Magneto-Therapy" described in James Randi's 1982 book, *Flim-Flam!* Randi encountered Wallace at a New Jersey "Psychic Fair," where she was selling small, powerful magnets wrapped in colorful plastic and labeled "North" and "South." Applied to the right place on the body, the magnets purportedly could bring about cures, and they sold well. (Of course, people who attend "Psychic Fairs" may not be as skeptical as most.) One of Sue Wallace's most interesting claims was that she could remove the toxicity of cigarettes by magnetizing them. She may be working for R. J. Reynolds by now.

The use of permanent magnets to enhance health has a long history in China and Japan and remains popular today. Several Asian companies continue to produce magnetic bracelets and necklaces reputed to cure assorted ills. A magnetic necklace said to cure headaches is accompanied by simple instructions: the necklace should be put on whenever the headache appears and removed as soon as it goes away. Since most headaches come and go, following these instructions precisely will clearly produce persuasive evidence of the necklace's power.

If magnets are strong enough, they have the ability to attract each other even when separated by, let's say, the thickness of an ear lobe. Thus another item of magnetic jewelry often marketed is magnetic earrings, each a pair of magnets (usually rare-earth magnets). Placed on opposite sides of the ear lobe, they hold without need for clips or piercing. And, some say, they are good for your health as well.

Nikken, a Japanese company formed in 1975, markets a variety of magnetic products, including pillows, mattresses, chair seats, and insoles for footwear (Magstep). Sales of their American branch, which is headquartered in Los Angeles, grew from 3 million dollars in 1989 to 40 million dollars in 1993. The permanent magnets in their products are arranged in a checkerboard pattern with alternating polarity, reportedly to assure that, in parts of the body in contact with the products, blood vessels in any and all directions will receive the beneficial effects of the magnetic fields.

One Japanese company markets small magnets mounted on circular bits of adhesive, "patch-on" magneto-therapeutic devices that they claim "stimulate blood circulation and relieve the stiffness in the shoulder, neck, and waist." Most of these employ ferrite magnets, but higher-priced versions with rare-earth magnets offer stronger magnetic fields. Several companies sell similar magnets specifically for application to acupuncture points, since many acupuncture practitioners and patients prefer magnets to needles. As explained by Larry Johnson in his *Magnetic Healing and Meditation*, "Magnet therapy has the advantage of being physically non-invasive. In this time of public concern for sterile technique, that is a big plus." Johnson reports that the correct "Magnet Mantra," or arrangement of the magnets on the body, can also greatly enhance the effects of meditation. Numerous testimonials on the book's cover report the benefits of this form of magnet therapy employed in conjuction with yoga, tai chi, and kung fu. It may not mean anything, but all of the individuals quoted have California addresses.

California has no monopoly on such products, though. A company only a few miles from my Massachusetts home distributes an impressive range of acupuncture and magneto-therapy items, all pictured in a 75-page catalog. It offers magnetic bracelets, necklaces, and earrings, magnetic belts, wrist bands, and knee wraps, magnetic vests, insoles, and seat cushions, a magnetic hammer ("for providing a relaxing percussion massage treatment"), and "water magnets . . . which may be attached to any

ANIMAL MAGNETISM

Figure 13.2 A different version of animal magnetism (from a recent *New Yorker* magazine). Perhaps the cow has recently swallowed a cow magnet (see next chapter) and the fish is a steelhead trout. (Drawing by Crawford; © 1993 The New Yorker Magazine, Inc.)

home or office water pipe . . . the natural way to treat hard or acid water." Ferrite, alnico, and rare-earth magnets of various sizes and shapes are also offered, including 3-inch-long alnico "cow magnets." (More about cow magnets in Chapter 14, but see Figure 13.2.) The company name is Oriental Medical Supplies (OMS), but the final page of the catalogue states, "Magnets are sold for personal or research use only. No medical claims are made." Can't be too careful in the medical supply business.

One Kentucky magneto-therapy company is much less careful. Its literature includes a four-page promotion for a book entitled *Curing Cancer with Supermagnets.* A photo shows a smiling patient who had been told that her cancer was incurable and she had only two months to live but who had been successfully cured by hanging a neodymium-iron-boron "supermagnet" around her neck. It is claimed that dozens of patients have been cured of lung, breast, and other cancers with this simple procedure. The miraculous effects of supermagnets are explained by the authors' "universal spiral theory," which is based on the observation that spirals are found throughout nature, from galaxies down to DNA. They propose that because the spirals in the supermagnets are bigger than the spirals in the cancer cells, the magnets eventually kill the cancer (nonsense, but scientific-sounding nonsense!). Their treatment with supermagnets is superior to surgery, radiation therapy, and chemotherapy, they claim, because these standard treatments can also damage healthy cells, which magnets do not. It's frightening to think that some misguided patients may forgo mainstream medical treatment in favor of being treated with supermagnets by superquacks. A man who sold magnets as a cure for cancer was recently arrested in Virginia, under a state law that forbids anyone but physicians and dentists from prescribing treatments for cancer. But *Curing Cancer with Supermagnets* is at least immune from that law, because one of the authors is a physician!

In *The Body Magnetic*, Dr. Buryl Payne discusses the past and present of magnetic therapy and speculates about its future, which may include "changing the vibration rate of the body so as to flip it into another dimension." (Even Mesmer didn't think of that!) Payne stresses the crucial importance of applying the proper pole of the magnet to the body, since south poles purportedly have a stimulating effect and north poles have a sedating or calming effect. It is therefore logical that tumors and infections should never be treated with south magnetic poles, a danger he warns about, in capital letters, in two different chapters.

If I thought there were any validity to the different physical effects of north and south poles, I would worry about people who sleep on Nikken magnetic mattresses and pillows, whose checkerboard pattern of alternating poles would therefore produce a pattern of alternating stimulation and sedation on their bodies. I further should warn the unwary that Payne and the OMS catalog denote the magnetic poles differently from mainstream science and technology. Apparently agreeing with William Gilbert (see Chapter 3), they define the north pole of a magnet as the end that points south. Those who experiment with magnetic therapy devices should read the directions carefully!

More cautious than most promoters of magnetic therapy, Payne writes, "Of course magnetic fields don't 'heal' anything; they only help the body heal itself. And they don't always help everything heal either. Magnetic field treatments are not a substitute for medical treatments, they are an adjunct and may be used in conjunction with any other dietary, chemical, or physical treatment."

Much of magnetic healing and magnetic therapy—of animal and mineral magnetism—was and is pure hokum. Yet many practitioners, Mesmer included, sincerely believed their claims, and many people have seen their health improved by magnetic treatments. The primary source of magnetic cures is surely psychological, the well-known "placebo effect," but so what? If it works, use it! (But in addition to, not instead of, traditional treatment.) The mind is, after all, part of the body, and it can contribute both to illness and to health.

When actual magnets are used, physical effects may also contribute to improvements in health in some cases. The electrochemical processes of the body are extremely complex and incompletely understood, and effects of magnetic fields cannot be totally ruled out. Real physical effects are a lot more likely from magnets than from crystals or pyramids, other objects attributed with magical powers by New Age gurus. The next chapter focuses on the use of magnets in the mainstream of today's medicine, and Chapter 15 considers magnetic fields created by the body and possible effects of magnetic fields on the body.

///// MEDICINE AND MRI

14

///// Magnets in Medicine

The preceding chapter dealt with approaches to healing that are normally considered part of "alternative" medicine. Here, as we consider the use of magnets in mainstream medicine, we must be fully aware that the borders between mainstream and alternative medicine are fuzzy and vary with time. In Mesmer's day, mainstream medicine prescribed bleeding and a variety of potions and poultices that often did more harm than good. (Franklin felt that some of Mesmer's cures may have resulted simply from keeping the patient away from conventional physicians.) Medical science has made tremendous progress in the two centuries since Mesmer, but today's mainstream medicine will seem pretty primitive 200 years from now.

One of the earliest uses of magnets in medicine was the removal of iron and steel objects from the body, such as shrapnel from unlucky soldiers and swallowed pins and nails from unlucky children. Both permanent magnets and electromagnets have been used, the latter having the advantage of on-off control. The device shown in Figure 14.1 extracting an open safety pin from the stomach can be turned off and on without passing

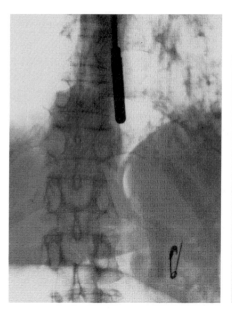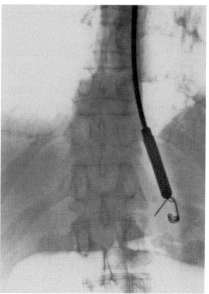

Figure 14.1 The removal of an open safety pin from a patient's stomach. A probe is "swallowed" by the patient *(left)* and maneuvered by the physician until the tip is near the rounded end of the pin. When the tip of the probe is magnetized (it is "turned on" when an alnico magnet in the body of the probe is pushed so that it touches the tip), it attracts the pin *(right)*. With the pin in this position, the point is less likely to do damage to the digestive tract as it is pulled out.

electromagnet currents into the patient's body. It has an iron tip and a moveable alnico permanent magnet inside. When the alnico is in contact with the tip, the tip becomes a temporary magnet and can attract steel objects (fact 5). When the alnico is pulled back from the tip, the tip is no longer magnetized. The doctor inserts the device and turns on the magnetic attraction only when the tip reaches the lower end of the pin. That way, the pin may be extracted with the sharp point directed down, minimizing the chance of damaging the esophagus or the stomach. The need to extract foreign objects from the digestive tract is fairly common with young children, who sometimes feel obliged to swallow anything small that they can get their hands on (except food).

Also rather indiscriminate in what it swallows is a grazing cow. Ingesting sharp steel objects can cause damage to the walls of the cow's intestines, a problem so common it has a technical name—"hardware disease." A popular preventive treatment is to encourage the cow to swallow a "cow magnet," mentioned previously. Although different materials and different shapes have been used, the common cow magnet is an alnico cylinder three inches long and a half-inch in diameter; it has rounded ends and is usually coated with a plastic for protection against corrosion or breakage. It remains in the cow's stomach (actually one of its several stomachs), attracting to it any steel objects the cow later swallows, such as bits of the wire used to bale hay. Although not a high-cost item, the magnet can easily be retrieved when the cow is slaughtered. (Prescribing cow magnets may not be mainstream medicine for most doctors, but it is for veterinarians.)

Back to human medicine: if you have a bone fracture that is slow to heal, you may find your orthopedist putting an electromagnet around the injured part. The coil in the "bone growth stimulator" is fed pulsed currents that generate time-varying magnetic fields that penetrate your body (fact 4) and induce electrical currents (fact 8) in the bone. The detailed mechanism for the effect is not known, but such currents usually stimulate bone growth, and PEMFs (pulsed electromagnetic fields) are FDA-approved for several orthopedic problems. Amorphous metals are sometimes used as core materials in the electromagnet (Figure 8.4). PEMF devices have been in clinical use for about twenty years, and over 100,000 fractures have been successfully treated, avoiding surgery that would otherwise have been necessary. The use of PEMFs in the treatment of osteoporosis and other disorders of the musculoskeletal system is under study. Induction of electrical currents within the body by localized time-varying magnetic fields produced by electromagnets has also been widely applied to the nervous system, including the brain, in research studies and specialized treatments.

One approach to treating cancerous growths is to kill the cancer cells by overheating, or "hyperthermia." Tiny capsules are filled with "soft" temporary-magnet materials with low Curie temperatures and delivered to the cancer site via a catheter. They are then exposed to AC magnetic fields from an electromagnet outside the body. Eddy currents induced in the magnetic material, which were a problem in transformers (chapter 8), are utilized here to raise the temperature of the tiny magnets. But the heating is limited to a specific temperature, the Curie temperature, at which the material in the capsule loses its magnetism (and the alternating magnetization and associated eddy currents therefore greatly decrease). Since the heating is both localized and limited, the damage to nearby healthy cells is minimized. Both amorphous metals and soft ferrites, with compositions chosen to produce Curie temperatures of about 110 degrees F, have been used. In some cases, anti-cancer agents have been incorporated into the capsules, which means that the capsules may be used for localized hyperthermia and chemotherapy.

Guidance of catheters through veins and arteries has been facilitated by the development of rare-earth permanent magnets. Tiny rare-earth magnets in the catheter tip, only one millimeter long and a half-millimeter in diameter, are easily guided by permanent magnets and electromagnets outside the body. This is another dramatic use of fact 4, magnetic forces acting at a distance, through intervening nonmagnetic material. Small rare-earth magnets, because of the substantial forces they provide, have also been useful in various dental and medical prostheses. For example, they may keep eyelids closed during sleep (in patients suffering from facial paralysis) or keep them open during waking hours (in patients with drooping eyelids caused by muscular dystrophy or other ailments).

Remember Dr. James Graham, who convinced many eighteenth-century Londoners that magnets were good for conception? Recently, a group of scientists from Georgia Tech have reported that magnets can be used for *contra*ception, in the form of magnetic IUDs (intrauterine devices). IUDs commonly have a string, called a "tail," that extends from the uterus into the vagina. The tail provides a mechanism for removal (the gynecologist retrieves the device by pulling on the string), as well as assurance that the IUD is in place (the woman can feel the tip of the string with her finger). Some studies suggest, however, that the string may also increase the chances of uterine infection by providing a path for microorganisms to travel into the uterus from the vagina. A magnetic IUD contains a small rare-earth magnet, the field from which can be detected from outside the body to confirm that the IUD is in place. The field also allows location of the IUD by an instrument (an extractor) that can remove the device with minimum discomfort to the patient. The magnetic IUD, being tail-less, is believed less likely to promote uterine infection.

Ultrafine magnetic particles, injected into the bloodstream and then localized in the body by exposing the patient to concentrated magnetic fields, have been used to deliver drugs to specific locations and to block (by "embolization") specific blood vessels. Magnetic particles to which antibodies have been attached will adhere to specific bacteria and can then be removed, along with the harmful bacteria, by magnetic forces. This is an example of "bioseparation," a technique that has become a powerful tool in the hands of microbiologists. Magnetic bioseparation has also been used to remove cancer cells from bone marrow, and by geneticists in DNA sequencing.

Magnets in the form of cyclotrons are used directly in some radiation therapy and to produce medical isotopes. One fascinating modern use of isotopes produced by medical cyclotrons is in positron emission tomography (PET). This medical imaging technique is based on the annihilation of antimatter by matter, which sounds like science fiction but is not. For example, bombarding the stable isotope nitrogen-15 with high-energy protons from the cyclotron produces oxygen-15, a radioactive isotope of oxygen that has a "half-life" of only 2 minutes. (Half of the oxygen-15 atoms produced will decay in 2 minutes, half of those remaining will decay in the next 2 minutes, and so on.) The oxygen-15 is combined with hydrogen to form radioactive water, which is injected into the bloodstream of the patient. In about a minute, the radioactive water travels to blood vessels throughout the body, including those in the organ you wish to study, commonly the brain.

When oxygen-15 decays, it emits a positron, a positively charged particle that is the antimatter version of the electron. Although stable in a vacuum, positrons don't last very long in ordinary matter like our brains (which, of course, are "ordinary matter" only from the physics point of view). The positron rapidly combines with an electron and the two disappear together, emitting two oppositely directed gamma rays. (In this process of antimatter-matter annihilation, mass is transmuted into pure energy.) A ring of detectors around the patient records the gamma rays, and a computer reconstructs the location of the positrons that emitted them, which is also the location of the oxygen-15, since the positrons don't get very far from their source. Thus the computer can form an image of the blood flow in the brain, plane by plane (a process called *tomography*—the *T* of PET). Several other positron-emitting isotopes have been used to study metabolic activity and other aspects of life processes.

Short-lived isotopes like oxygen-15 are desirable for such studies, because they emit positrons quickly and therefore produce high counting rates in the detectors. Furthermore, their rapid decay means that they soon are harmless to the patient. But isotopes with short half-lives have to be used soon after they are produced, and thus the cyclotrons must be at, or very near, the hospitals. Fortunately, the higher fields and higher current densities available with superconducting magnets have greatly reduced the size and weight of cyclotrons capable of producing particles of a given energy. If these cyclotrons had to be built with copper-wound electromagnets, like the original cyclotrons of Lawrence, they would be nearly twenty times as heavy.

There are specialized uses of magnets in medicine beyond those mentioned above. But the one application of magnets that has had by far the biggest impact on modern medicine is MRI—magnetic resonance imaging.

///// Personal Images

MRI is such an important and exciting application of magnets that it's worth a bit more explanation than the devices mentioned so far. MRI is one of several methods doctors now have to study the insides of the human body without cutting it open, and the images produced are remarkable (Figure 14.2). Let's first focus on imaging in general—the *I* of MRI.

If you look at a photo in your local newspaper through a magnifying glass, you see that the picture actually consists of many tiny dots arranged in a square pattern. In light areas of the photo, the dot in each picture element or *pixel* is small and there is lots of white space between neighboring dots. In the dark areas, the dots are big enough to overlap and little or no white space is left. Without the magnifying glass, and with the paper held at normal reading distance, we no longer see the individual dots (at least not at my age). The varying sizes of the dots in each pixel produce various tones of gray, producing an image that appears continuous. But we know it is not continuous; it actually consists of an array of tiny pixels of varying intensity.

Magnetic resonance images and other images of our insides, such as PET scans and X-ray CAT (computerized axial tomography) scans, also consist of an array of pixels of different intensity.

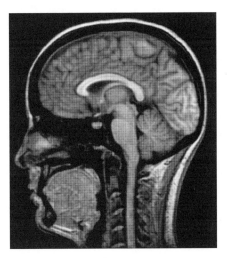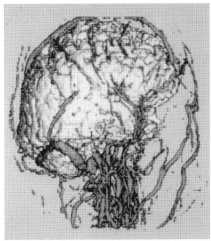

Figure 14.2 Two views of the brain. A two-dimensional NMR image *(left)* represents a single plane, or "slice," through the head. A three-dimensional image of the brain and blood vessels in the head (facing to the right) may be generated by computer *(right)* from a series of two-dimensional images. Two-dimensional images are commonly used for diagnosis, but three-dimensional images can be useful to physicians in planning surgery.

In these images, each pixel corresponds to a small piece of the body at a particular location, and the intensity of each pixel corresponds to some physical property of that piece of the body.

The first of these imaging techniques to be developed was the X-ray CAT scan (or CT scan). Standard X-ray pictures, like those your dentist takes of your teeth, are made simply by sending x-rays through your body in one direction and using a film to detect how many get through. The intensity of X-rays hitting each part of the film is determined by the total amount of X-rays absorbed along the entire path through the body, not those absorbed by one particular location. If X-ray beams are instead passed through the body in a large number of different directions, however, and the intensity of each beam is then detected, a computer can use the data to calculate the X-rays absorbed in each location. Images of a cross-section of the body can then be produced by electronically translating the computer's results into an array of pixels on a film or screen, with the intensity of each pixel corresponding to the X-ray absorption at a specific location in the body.

Producing body images by CAT scan is clearly much more complex than producing traditional X-ray images. Instead of a single photographic film, a large array of detectors is required, and the resulting data require complex computer calculations to produce an image. But CAT scans are so much more useful that they rapidly became an important tool for medical diagnosis, and two early developers of the system, Allan Cormack and Godfrey Hounsfield, were awarded the 1979 Nobel Prize for medicine.

With positron emission tomography, the computer calculations are very similar to those used in CAT. In PET, however, the detectors count the gamma rays emitted whenever positrons are annihilated by combining with electrons. The intensity of each pixel in a PET scan corresponds to the number of positrons emitted at each specific location in the body (since, as noted earlier, they are not able to travel far from where they are emitted). The number of positrons emitted in turn is determined by the concentration, at that location, of the radioactive isotope that decays and emits the positrons (oxygen-15, in the example given earlier).

As with CAT and PET, the intensity of each pixel in a magnetic resonance image corresponds to specific physical properties of a specific location of the body. The local physical property recorded in a CAT scan is X-ray absorption, and the local physical property recorded in a PET scan is the concentration of a positron-emitting radioactive isotope. To understand the local physical property represented by each pixel in an MRI scan, we must first discuss MR—magnetic resonance—and the letter that was discreetly removed from the name of this imaging technique years ago: *N* for nuclear.

///// Nuclear Magnets

MRI is really NMRI—nuclear magnetic resonance imaging. But the public at large has a negative reaction to the word *nuclear*, which is associated in some minds with the horrors of nuclear bombs and the unpopularity of nuclear power. So marketers of medical imaging systems based on NMR soon dropped the *N*. Fair enough. Whereas nuclear bombs and nuclear reactors involve, in *vastly* different degrees, nuclear reactions that produce dangerous particles, NMR involves only the magnetic properties of atomic nuclei. Unless the magnetic fields involved in MRI have some harmful effects not yet identified (see next chapter), MRI is far safer to the human body than CAT or PET, since both X-rays and gamma rays damage bodily tissue (yet CAT and PET are benign acronyms, at least to people who like domestic animals).

It wasn't until early in this century that we learned much about the size and internal structure of atoms. The diameter of an atom is only a few tenths of a nanometer; a million atoms in a row would barely reach across the period at the end of this sentence. Most of the volume of an atom is taken up by electrons, the elementary particles of negative electric charge. The atom's nucleus, home for the positively charged protons (and, in most atoms, some neutrons), occupies only an infinitesimal fraction of the atomic volume. (Picture an atom scaled up until its diameter equaled the width of the continental United States. On that scale, the nucleus would be about the size of a farmhouse in the middle of Kansas—like Dorothy's.) Although most of the volume of an atom is taken up by electrons, most of its weight is in the nucleus, since protons and neutrons are each nearly two thousand times heavier than electrons.

Electrons are the source of most magnetism, as discussed in some detail in Chapter 6. But the nuclei of many atoms, including the nuclei of hydrogen, are weakly magnetic. We have lots of hydrogen atoms in our bodies, mostly in water molecules, and NMR images are usually produced from magnetic resonance of hydrogen nuclei, the simplest of all atomic nuclei. (Hydrogen atoms in molecules of fat also make important contributions to NMR images. Even if your body is not fat, it contains lots of fat molecules!)

The nuclei of most hydrogen atoms are simply lone protons, tiny magnets one femtometer (10^{-15} m) in diameter, and their magnetism is caused by the spinning motion of the proton about a central axis. The magnetic strength of a proton is pretty puny, less than a thousandth of the magnetic strength of an electron. There are 66 billion billion (6.6×10^{19}) protons in a cubic millimeter of water, but their magnetic poles usually point randomly, in all directions, and therefore give no net magnetism. Put these spinning protons in a large magnetic field, however, and things change.

Imagine yourself lying on your back inside an MRI magnet, a giant superconducting magnet that produces a magnetic field directed along the length of your body. The magnetic field is about 20,000 times stronger than the earth's magnetic field, and you don't feel a thing. But those little magnetic protons do.

Like tiny compass needles, the protons try to align themselves along the field direction, but to do so they must overcome the random jiggling associated with temperature that tends to keep them pointing at random. It's a tough battle, but the number of proton magnets pointed along the field becomes just a little higher than the number pointed in the opposite direction, giving a small net nuclear magnetization in the field direction. The net imbalance is small: in a field of 10,000 gauss, it corresponds to only about 3 protons per million. But that's still over 200 trillion (2×10^{14}) protons in that cubic millimeter of water, and that net imbalance is what is recorded in MRI.

There are two ways we could increase the imbalance and get

a larger net nuclear magnetization. We could increase the field, which is trying to align the tiny magnets, or we could decrease the temperature, which is trying to randomize them. We can't lower the temperature very much and still keep you alive in there, so it's desirable to use a magnetic field as large as possible. That's why most MRI systems use superconducting magnets.

///// Magnetic Resonance

If the axis of a spinning top or gyroscope is not exactly vertical, it wobbles around in a regular motion that physicists call *precession*. If the arrow representing the *net* magnetization of hydrogen nuclei is not exactly parallel to the 10,000-gauss field that produced it, the arrow precesses around the field direction (Figure 14.3A), much as the axis of a top precesses around the vertical. The protons, after all, are spinning like tiny tops, and that spin is associated with their magnetism. The rate at which the net magnetization precesses depends both on the magnetic strength of the nucleus, in this case a lone proton, and on the strength of the magnetic field. In a field of 10,000 gauss, protons precess 42.57 million times each second—at a frequency of 42.57 MHz. That's in the range of radio frequencies (RF).

Suppose we now install another electromagnet that produces a small magnetic field directed *across* your body, perpendicular to the 10,000 gauss field. But suppose that instead of a steady current, we apply RF currents to this coil so that this small transverse magnetic field alternates at exactly 42.57 MHz. This transverse field exerts only a small alternating force on your precessing protons, but the alternations of that force are synchronized exactly with the precession frequency. As anyone knows who has swung higher and higher in a backyard swing with the use of well-timed "pumping" (or aided by a convenient friend providing well-timed pushing), even a small force exerted at the

Figure 14.3 Net nuclear magnetization of the atoms in a human body during magnetic resonance imaging. (A) Before the radio-frequency (RF) field is applied, the net nuclear magnetization (represented here by a bold arrow) is nearly parallel to, but precesses around, the static field. (B) After brief application of the RF field, the net nuclear magnetization swings around in a plane perpendicular to the static field; the changing magnetic field induces a current in the RF coil. (C) The current induced in the RF coil changes over time. The initial signal amplitude indicates the strength of the net nuclear magnetization, which is a measure of the local mobile proton density; the rate of decay of the amplitude yields the local relaxation time. By adjusting various parameters, the operator can change the relative weighting of these two variables—proton density and relaxation time (as a function of position)—and thereby adjust the quality of the images that are developed.

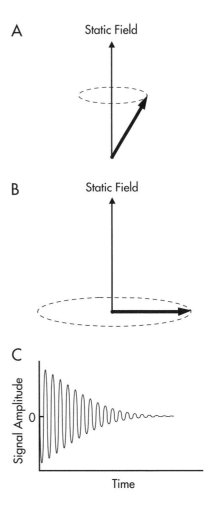

right frequency—a *resonant* frequency—can produce a substantial cumulative effect. The transverse RF magnetic field, oscillating at a frequency in resonance with the precession frequency of the protons, moves the net nuclear magnetization further and further away from the direction of the 10,000-gauss static field with each cycle. A small driving force exerted in synchronized fashion 42.57 million times per second—at exactly the same frequency that the magnetization is precessing, so that the force from the RF field is applied at the same point in each precession cycle—can overcome the large force exerted by a much stronger field. That's the wonder of magnetic resonance.

If the RF field is turned off just as the net nuclear magnetization reaches the plane perpendicular to the static field (the field along your body produced by the superconducting magnet), the magnetization temporarily swings around in this plane (Figure 14.3B). In this position, the precessing magnetization produces an alternating field of maximum strength in the transverse direction, which can be detected, using fact 8, by the same coil that stirred things up in the first place. With the RF field turned off, the magnetic orientation of the protons will gradually swing back toward its position in Figure 14.3A (where it was before the RF field was turned on), but it takes some time, soothingly called the *relaxation time,* to return. The current induced in the RF coil will look something like Figure 14.3C, and a well-instructed computer can record both this relaxation time and the initial amplitude of the alternating currents, which will tell us something about how many protons produced the signal.

I find this amazing. The magnetism from the electrons in your body is very, very weak, and the magnetism of your nuclei is much weaker still. Yet with a superconducting electromagnet producing a field of 10,000 gauss to partially align your nuclear magnets, and with another electromagnet delivering RF power at just the right frequency to those magnets, super-sensitive electronic equipment can detect this weak nuclear magnetism. This is not at all what Mesmer had in mind, but NMR has detected a bit of your "animal magnetism."

With the important help of a large superconducting magnet, the RF coils have produced and detected NMR, the resonance of nuclear magnets within your body. But to produce an image, we somehow have to separate the NMR signals coming from differ-

parts of your body. In 1973, a professor at the State University ⌐ New York at Stony Brook named Paul Lauterbur suggested that we could produce images if we added more coils.

///// The Imaging Technology

Lauterbur's idea was to add "gradient coils" that added or sub-tracted small fields to the main 10,000-gauss field in such a way that these changes were different at different locations in your body. Now, instead of the precession frequency of your protons being exactly 42.57 MHz at every point in your body, the protons would precess at slightly different frequencies at different points in your body because of slight changes in the field from point to point. With three gradient coils added (one for each direction in space) and the RF coil and the various gradient coils pulsing in a special time sequence controlled by a computer, signals de-tected by the RF coil can be analyzed by the computer to calculate "mobile proton" densities and relaxation times for each location in your body. (The NMR signal comes from the nuclei of hydro-gen atoms in liquids and soft tissues, atoms that are relatively mobile. Protons in solid tissues like teeth and bone have such short relaxation times that they are not detected.) These data can then be combined in various ways to define the intensity of a pixel corresponding to each location in your body, and, as in a newspaper photo and in CAT and PET scans, the pixels of vary-ing intensity are arranged to yield an image.

In CAT and PET imaging, the intensity of each pixel corre-sponds to only one physical property: X-ray absorption in the former, isotope concentration in the latter. In NMR imaging, both mobile proton density and relaxation times influence the signal. The operator can use minor changes in the timing of the various pulsed fields to change the relative importance of proton density and relaxation times in determining the intensity of each pixel, thereby producing different degrees of contrast in the image. For example, relaxation times in many tumors are much longer than in healthy tissue, and images that emphasize relaxation time differences show such tumors very clearly. Because of this good contrast, because bone does not contribute to the images, and because the position and orientation of the scanning plane can be varied electronically without moving the patient, MRI is an excellent method of studying brain tumors.

More generally, MRI yields better contrast between different soft-tissue structures than CAT scans. It has been found to be especially powerful in studying the spinal cord, joints, and other parts of the musculoskeletal system as well as blood vessels and blood flow. Although each imaging system has its strengths and weaknesses, the versatility and noninvasive nature of MRI, plus the absence of tissue-damaging X-rays or gamma rays, has led to widespread use of MRI in medical diagnosis. If only it weren't so expensive . . .

///// Magnets for MRI

An MRI examination can easily cost a thousand dollars, largely because the equipment cost the hospital about 2 million. Much of the cost is in the superconducting magnet, which is large enough to admit a human body and strong enough to produce a field of many thousands of gauss. The development of high-field superconductors in the 1960s made this possible.

The conductor commonly used for the main magnet contains many fine filaments of a superconducting niobium-titanium alloy imbedded within a matrix of copper. Producing such a complex conductor requires a series of sophisticated metallurgical processing steps, which is expensive. The conductors must then be carefully wound into large electromagnet coils, usually six separate coils, designed to produce a very homogeneous field over a very large volume. Further fine-tuning is done with separate "shim coils" to produce a magnetic field that is uniform to a few parts per million. Three gradient coils and an RF coil are also necessary parts of the system. The superconductor must be cooled with liquid helium, which must be contained within a thermally insulated container, called a cryostat. Pulsing the RF and gradient coils and extracting the data to create images require complex electronics and computer software. No wonder MRI is expensive.

Although most MRI is done with superconducting magnets, systems based on permanent magnets, such as neodymium-iron-boron magnets, are also in use (Figure 14.4). The magnetic fields of the rare-earth permanent magnets are enhanced with iron pole pieces, much as in the sixteenth-century armed lodestones dis-

Figure 14.4 Sumitomo 24-ton magnet assembly for MRI. At top and bottom are large disks of neodymium-iron-boron permanent magnet. The remainder of the assembly is iron or steel and provides a return path for the magnetic field. In the central region, there is a vertical magnetic field of 2 kilogauss.

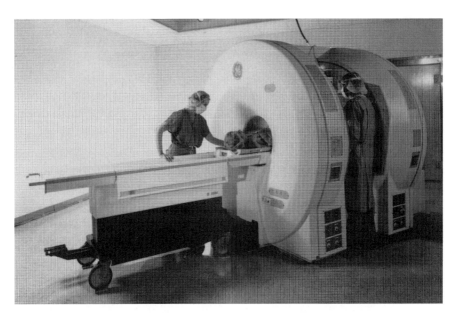

Figure 14.5 New split-coil system allows physician access to patient and MRI scans during surgery.

cussed in Chapter 2. The field strength is 2,000 gauss and is vertical in this design, rather than horizontal as in the superconducting system. Gradient and RF coils are not shown.

The lower magnetic fields available with permanent-magnet MRI result in some decrease in the quality of the images, but because of their simplicity (no liquid helium required!), compactness, and lower cost, several hundred systems have been installed around the world. Most are manufactured by Sumitomo, the world's leading producer of neodymium-iron-boron. Each system requires over two tons of permanent magnets, making MRI systems one of the highest-volume applications of neodymium-iron-boron.

An exciting new system has recently been introduced by General Electric (Figure 14.5). In conventional MRI, the superconducting magnet surrounds the patient, blocking physician access.

Images are used for diagnosis only. The new design splits the superconducting magnet in half, allowing the physician access to the patient during imaging. It also is far less frightening to claustrophobic patients than the conventional design.

Clinical trials began at Brigham and Women's Hospital in Boston in March 1994. Doctors will first explore the system's use in several forms of minimally invasive surgery, such as needle biopsies and tumor treatments using laser surgery or cryosurgery (freezing). By generating NMR images during an operation, the system provides much more information about the patient's anatomic structures than is normally available to the surgeon. Progress of the surgery can be monitored as it is performed.

In addition to the innovation of a split-magnet design, this system employs several other technological advances. Rather than niobium-titanium, the superconductor used is a compound of niobium and tin. Because this compound has a higher critical temperature than niobium-titanium (18 versus 10 degrees K), the magnet may be operated at a higher temperature and cooled by gaseous, rather than liquid, helium. In addition, the electrical leads connecting the niobium-tin to the outside electrical supply are made from one of the recently discovered ceramic "high-temperature" superconductors—yttrium barium copper oxide. The current-carrying ability of this superconducting oxide is enhanced by the application of a large magnetic field to a slurry of powdered ceramic before the powder is heated to form a dense solid. (This process was one of the last projects I worked on at General Electric before moving to MIT in 1989.) Current leads to superconducting magnets have become one of the first industrial applications of the new oxide superconductors.

In most MRI systems, strong, steady, and invisible magnetic fields extend well beyond the magnet itself. These fields, much stronger than anyone is likely to encounter anywhere else, can be both useful and very dangerous.

I have personally found these fields useful in demonstrating to students, in a dramatic way, some basic aspects of magnetism, particularly the generation of eddy currents. A thin metal sheet balanced on one edge near the end of an MRI magnet will tip only very slowly, as if falling through molasses. As the sheet tips from perpendicular to the field to parallel to the field, the total field penetrating the sheet decreases. This change induces eddy currents (fact 8) that oppose the motion. Similarly, a large metal ring can be turned from perpendicular to the field to parallel only with some effort. In each case, the motion of the metal changes the field penetrating it, inducing eddy currents that oppose the motion. Students can easily see and feel fact 8.

Before entering the magnet room, however, students are instructed to leave their wallets (which may contain credit cards), watches (some of which may malfunction if a critical steel part becomes magnetized), and any loose steel and iron objects, even hairpins, outside the room. Any magnetic objects can be drawn suddenly into the magnet with great force and speed, causing damage to anyone who may be in the way. Medical personnel and others working with MRI are well aware of these dangers.

Obviously patients with magnetic implants within their bodies from prior operations may not be good candidates for MRI. Neither are patients with cardiac pacemakers, which may malfunction or become permanently damaged from the strong magnetic fields. Even nonmagnetic metals in the body may be a concern, since the pulsed fields of NMR imaging will induce electric currents in any metal. Although such currents may not be harmful to the patient, they may produce artifacts in the image. Stainless steels in dentures and orthodontic braces are known to obscure images of the facial region.

Even for patients without pacemakers or metallic implants, questions have been raised about possible harmful effects of the strong steady fields and various pulsed fields of MRI on the human body. Recent newspaper articles have suggested that even the much weaker magnetic fields from electric power lines and home appliances may have harmful effects. In the next chapter I discuss these controversial questions, but first I describe some remarkable recent advances in detecting the magnetic fields created by the human body itself, including fields generated by electrical activity in the brain.

///// BIOMAGNETISM

15

///// Living Magnets

Your body, like any matter, has a magnetic response when exposed to external magnetic fields. If you are examined by MRI, for example, the strong magnetic field from the superconducting magnet produces a net nuclear magnetization within your body, and the RF magnetic field, if its frequency matches the precession frequency of that magnetization, elicits a response (Figure 14.3) that is mapped as the NMR image. Other areas of medical study also take advantage of the magnetic response of the body to external magnetic fields. Most of the iron in your body, as noted earlier, is in your blood, but much of it is also stored in the liver within a protein called *ferritin*. In some diseases, the iron content in the liver is much reduced (iron deficiency) and in others it is increased (iron overload). Sensitive field detectors that measure the body's magnetic response to applied magnetic fields can be used to diagnose these conditions. Noninvasive magnetic measurements are also very useful in the search for fine magnetic particles that accumulate in the lungs, especially in the lungs of welders, machinists, and foundry workers.

As interesting and important as these aspects of biomagnetism are, they relate only indirectly to the nature of a human being as a living organism. Most NMR images of your body would look pretty much the same whether you were alive or dead. Measurements of the iron in ferritin stored in your liver, or of the magnetic particles trapped in your lungs, would give pretty much the same results in dead bodies as in live ones. But your body is more interesting when you are alive, even magnetically. Your life processes produce weak magnetic fields of their own, which today can be measured. In 1963, Gerhard Baule and Richard McFee took their experimental equipment to a rural area outside Syracuse, New York, and recorded the magnetic fields produced by a beating heart. They were the first to measure directly the magnetism of life.

The action of muscles and nerves within the body involves the motion of ions (charged atoms), and the electric current produced by the movement of charged particles creates a magnetic field (fact 6). We are all electromagnets. But we are very, very weak ones, which makes things tough for the biomagneticians trying to measure these weak fields. The magnetic field outside the chest created by a beating heart is less than a microgauss. This is a fluctuating field, but it is weaker than fluctuations in the earth's magnetic field, and weaker than the magnetic fields created by "urban noise" from motors, cars, power lines, and other features of our electromagnetic society. The fields produced by the devices in our built environment are typically a few milligauss, hundreds of times stronger than even the strongest biomagnetic fields.

How could Baule and McFee measure the weak biomagnetic field of the heart in the midst of larger, environmental magnetic fields? First, they left the electrically active environs of urban Syracuse for a rural area, to reduce the background "noise." Second, instead of using a single coil of wire to measure the heart's field (by the currents induced in the coil, via fact 8), they used two identical coils, wound in opposite directions, placed at different distances from the chest. Magnetic fields from distant sources (such as a power line) would be expected to be very nearly the same at the two coils and to induce equal but *opposite* currents, which would cancel. But the field from the heart should be higher at the coil nearer the chest, the two induced currents would not fully cancel, and the electronics should detect a difference between them. This two-coil system, called a *gradiometer*, has been incorporated in the much more sensitive detecting systems used today.

The next advance was made by David Cohen of the University of Illinois, who decided to build a small room in his laboratory shielded from the magnetic fields of his Chicago environment. (That's easier than lugging all the experimental equipment out to the country!) Cohen's room was walled with a soft magnetic alloy, to attract DC and low-frequency fields and thereby keep them from inside the room, and with aluminum, to keep out high-frequency magnetic fields via induced eddy currents. When his magnetically shielded room was being built in 1965, Cohen was disturbed to learn that an electromagnet producing a large time-varying magnetic field was being installed in the laboratory next to his, and he complained to university officials. He was told that biomagnetism was absurd and that he would never detect anything anyway.

The magnet next door stayed, but Cohen's measurements in the shielded room proved the officials wrong. He verified the Baule-McFee measurements of the heart's magnetic field in 1967, and soon thereafter, with time-averaging of electrical signals to further minimize the interference from external fields, he was able to detect even the much weaker biomagnetic fields from the brain and other organs. His exciting results began to be noticed, and MIT beckoned. Cohen moved to Cambridge, Massachusetts, where he built a more elaborate shielded room. But his experimental techniques were still cumbersome, and one more advance was needed to ignite widespread interest in biomagnetic fields.

That advance was provided by Ford scientist James Zimmerman and his SQUID.

The SQUID, or Superconducting QUantum Interference Device, is capable of measuring extremely weak magnetic fields. Zimmerman was fascinated with his SQUID and spent several years improving it, but he hadn't yet zeroed in on an important application for it. (His project must have severely tested the patience of his managers at Ford.) The SQUID was so sensitive that it easily detected the motion of metal chairs yards away, an effect which his lab colleagues enjoyed demonstrating. Zimmerman said wryly, "It was obvious that we had an extremely sensitive detector of lab chairs."

In 1969, Cohen, with his interest in detecting weak biomagnetic fields, and Zimmerman, with his "lab chair detector," got together. To coin a phrase, the rest is history. Shortly after Christmas, Zimmerman's SQUID was installed in Cohen's shielded room (Figure 15.1). Clad only in undershorts, Zimmerman climbed into the room and sat with the SQUID magnetometer at his chest. Cohen adjusted the equipment, and soon Zimmerman's heartbeat was recorded with the clarity of an electrocardiogram. The article describing their results, published in April 1970 in the *Journal of Applied Physics,* has been called "the Magna Carta of a new field: biomagnetism." Two years later, Cohen reported in *Science* the first SQUID measurements of the magnetic fields of the brain. These two seminal papers triggered worldwide interest and a growth in research activity that continues to this day.

Experimental techniques for measuring biomagnetic fields are much more sophisticated today than they were in 1969. Complex gradiometers linked to SQUIDs can measure weak biomagnetic fields without the need for magnetically shielded rooms (although shielded rooms still help). Devices with more than 100 SQUIDs, like the one in Figure 15.2, now can simultaneously record biomagnetic fields from large areas of your head or body. SQUID arrays are used today in many laboratories to study

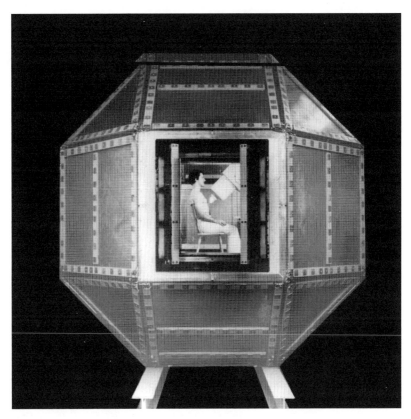

Figure 15.1 In a magnetically shielded room at MIT, a SQUID is positioned to measure the magnetic field from a human heart.

the most interesting and most complicated organ in the human body—the brain.

///// The Magnetic Mind

VCRs, computers, and MRI systems are all pretty complicated, but compared to the human brain they are simple. In the outermost layer of the brain, the cerebral cortex, there are over 10 billion neurons (nerve cells) linked via over 100 trillion synapses

Figure 15.2 Whole-cortex magnetoencephelography (MEG) helmet contains over 100 SQUIDs. The device measures the distribution of magnetic fields from brain activity.

(interconnections). There is still a lot that we don't understand about the complex electrochemistry that goes on in the individual neurons and synapses, but the vast communication network they form is much more of a mystery. What goes on in there when we see, hear, smell, taste, and touch? When we stand, walk, run, hit a tennis ball? When we fear, hate, love? When we learn, remember, think, and even think about thinking? One of the newest means that scientists now have to explore the mysteries of the mind is MEG, or magnetoencephalography (a term introduced by Cohen in 1972), the measurement of the tiny magnetic fields (a few nanogauss) generated by the brain.

For over 60 years, electrical activity of the brain has been studied by EEG (electroencephalography), which employs electrode pairs attached to the scalp. EEG has been useful in the diagnosis of epilepsy, brain tumors, cerebral infections, and serious head injuries. It has also been used in "biofeedback" techniques, such as relaxation and meditation. EEG now has a powerful partner.

One goal of both EEG and MEG studies is to identify the location in the cortex of the electrical currents associated with various brain functions—the processes responsible for perception, movement, emotion, and complex cognitive tasks. MEG has been found to be a bit more precise than EEG in localizing neural activity. More important, MEG is sensitive primarily to brain currents flowing parallel to the skull, whereas EEG detects both currents flowing parallel to and currents flowing perpendicular to the skull. A combination of the two techniques therefore allows better separation of the different components of the brain's electrical activity. Although PET and MRI yield more (and different) information about the deep interior of the brain, they produce images relatively slowly. EEG and MEG are fast enough to keep up with the rapid motion of brain activities around the cortex and so are more capable of studying the dynamics of the brain.

MEG research is now carried out at many laboratories around the world, with major efforts in Finland, Germany, Japan, Canada, and the United States. A group at New York University, for example, has been mapping the magnetic fields of the brain for over twenty years. One focus of the NYU group has been to locate specific sites in the brain where the alpha rhythm (a particular pattern of brain activity) is briefly suppressed for specific thought processes—for example, when visual images are suggested ("imagine a boat")—in an individual whose head is surrounded by a set of SQUIDs. The subject sits within a small room magnetically shielded from electromagnetic noise, a large part of which originates from the BMT subway line twelve stories below the laboratory. Biomagnetic research can flourish even in the "urban noise" of downtown Manhattan.

Another noninvasive approach to functional mapping of the brain is localized magnetic stimulation. A pulse of electric current is delivered to a figure-8 coil positioned outside the head. This produces a time-varying magnetic field, focused at the cross point of the figure eight, that stimulates localized brain currents (via fact 8). The resulting bodily response yields information about the function of the particular area of the brain exposed to the field. The recent development of MEG, MRI, PET, and magnetic stimulation techniques, in coordination with long-established EEG techniques, have given specialists a powerful assembly of approaches with which to analyze the most complex organ of our bodies.

In addition to its uses in research, MEGs have already found clinical applications. They have been used, in coordination with EEGs and other brain-imaging techniques, to prepare surgeons for various forms of brain surgery, including treatment of epilepsy and brain tumors. As experience and understanding of this relatively new technique grow, physicians will find more and more reasons to measure the magnetic fields of the mind.

///// Killer Gauss?

Our bodies as electromagnets generate biomagnetic fields of very low strength (nanogauss to microgauss). Our bodies are exposed, however, to stronger magnetic fields: those on the order of milligauss, from our high-tech environment, to nearly a gauss, from the earth's "deeply hidden" electromagnet. Furthermore, in various alternative and mainstream medical treatments, we may be exposed to fields that may reach many kilogauss. What effects these magnetic fields may have on us and on other living organisms is today a highly controversial question. (This area of research is sometimes called *magnetobiology*, whereas the term *biomagnetism* is commonly reserved for studies of the magnetic fields *generated by* living organisms.)

Some people are convinced that the magnetic fields we receive from electric power lines, electric appliances, computer terminals, and other components of twentieth-century technology increase the incidence of cancer. (This view might be called the "milligauss as killer gauss" perspective.) Alarmist books promoting this view, with titles like *Currents of Death,* have created widespread public concern. Other people consider the evidence supporting such views completely unconvincing and present scientific arguments that the small fields we normally encounter could not possibly have deleterious effects. Although I lean toward the latter view, I must admit that I'm not sure that magnetic fields in today's environment are totally harmless.

Let's start with those large steady fields in MRI magnets. Fields up to 20 kilogauss are in use and approved by the FDA, systems producing 40 kilogauss have been built and tested, and 100-kilogauss systems have been proposed. In one 40-kilogauss study, proton images were first obtained on two anesthetized dogs, and there was no subsequent evidence of any injury to the dogs. Eighteen human volunteers were then tested, with exposures to the 40-kilogauss field varying from less than an hour to a total exposure of over 30 hours (over many sessions). The subjects' vital signs were monitored and other tests—urinalysis, blood chemistry studies, ECGs, EEGs, and psychological tests—were done. No harmful health effects on any of the volunteers was detected.

On the other hand, several of the volunteers, and workers involved with the magnet installation, reported a variety of sensory effects. Some reported dizziness, nausea, and brief difficulty maintaining balance when they left the magnet (as one experiences when getting off a merry-go-round). Some reported a peculiar metallic taste in their mouth, and a few reported a visual perception of flashing lights (usually noticeable only when the room was darkened). This latter effect correlated with rapid head motions while the subjects were still in the magnetic field, and is presumably related to the phenomenon known as "magneto-phosphenes," observed over a century ago when AC electromagnets of 10–100 Hz were placed near the head. The flashing lights, and perhaps most of the other sensory effects, result from electric currents induced (fact 8!) in the body by changing magnetic fields or by the body's own movement within a steady high field. (As noted earlier, electric currents are basic to the operation of the nervous system, so it is not surprising that induced electric currents could have sensory effects.)

Even higher steady fields have been used in studies on other organisms. Mice, fruit flies, sea urchin eggs, and bacteria have been exposed to fields up to 140 kilogauss at the Naval Research Laboratory in Washington for 1 to 2 hours. All survived with no observed ill effects. In studies in Grenoble, France, quail eggs exposed to fields over 60 kilogauss for up to 16 days developed and hatched normally. Eleven rats exposed to 100 kilogauss for 8 hours a day during pregnancy had healthy litters, with no significant differences observed between the experimental group and a control group.

Despite these and other studies reporting no significant observable effects of exposure to extremely high steady fields, other experimenters have reported effects on living systems even in much lower fields. Some studies may have been influenced by subtle factors unrecognized by the experimenters. For example, increases in skin temperature of mice, pigeons, and humans after exposure to magnetic fields now appear to have been caused by changes in the convective motion of air produced by weak magnetic forces on oxygen molecules in the air, rather than by any magnetobiological effects. But many reports remain of effects of magnetic fields on biological systems, both on the molecular level and on complete organisms, and not all have been explained away.

The current public controversy centers not on high steady fields but on weak AC magnetic fields of power-line frequencies (60 Hz in the United States). In comparison to RF and other high-frequency fields that our bodies are exposed to every day, 60-Hz fields are called ELF (extremely low frequency), and sometimes ELF-EMF (extremely low frequency electromagnetic fields). Here the research that has caused most concern is not laboratory studies on biological systems but epidemiological studies reportedly showing a statistical link between some forms of cancer and proximity to electric power lines or high-field exposure in the workplace. One of the most publicized studies was a 1992 Swedish report of a doubling of leukemia risk for children living in houses with ELF magnetic fields of 2 milligauss or more. These results have been criticized on several grounds, but they drew considerable attention because they seemed to support several earlier, but less scientifically rigorous, studies.

One of the most convincing arguments against a connection between ELF and leukemia is that per capita residential use of electric power has increased twenty-fold in the last fifty years without any significant change in the incidence of leukemia. (Overall death rates from cancer have increased over that time period, but mostly from increases in respiratory cancer, presumably caused by smoking.)

One argument against the possibility of ELF hazards that I find less convincing is that no realistic physical mechanism has yet been identified that could explain harmful effects from weak 60-Hz fields. (ELF magnetic fields can stimulate nerves and heat human tissues, but only at fields much higher than milligauss.) Our current understanding of the electrochemistry of the human body, and of the factors that encourage cancer, is very limited. Harmful effects of weak ELF fields, like many other things in this complex world, may occur even if we do not understand exactly how they are caused. One possibility recently suggested by Joseph Kirschvink of Caltech is that fine ferromagnetic particles, located at special locations within our body, could amplify the effects of low magnetic fields and, for example, thereby modify ion flow across cell membranes.

In 1892, two researchers who had failed to find any effects of strong magnetic fields on humans concluded their study with puzzlement: "magnetism is certainly a remarkable force, and we find it very difficult to understand why it seems to have no influence whatever upon the human body and its wonderfully delicate neuroelectric mechanism." I agree. It seems likely that magnetic fields much stronger than the earth's magnetic field could affect our biological processes, sometimes in a helpful manner, sometimes harmful.

It seems less likely that fluctuating magnetic fields with an amplitude much weaker than the earth's magnetic field could have strong effects. Thus I also agree with a 1992 paper that concluded: "The average person has nothing to fear about sleeping next to an electric radio/alarm, using household appliances, or walking under high-tension power lines. He or she should, however, think twice about lighting up and look both ways before crossing the street." Although there is ample evidence that cigarettes and automobiles are serious health hazards, the evidence produced to date about the dangers of ELF fields remains unconvincing. Yet I must admit that nowadays I walk a little faster when I pass under the high-voltage line in the woods near my house. And I no longer sleep under an electric blanket.

///// The Birds and Bees

Cole Porter wrote, "Birds do it, bees do it, even educated fleas do it." Evidence is not yet in on educated fleas, but many scientists believe that birds and bees, and many other animals as well, have magnetic sensors within their bodies that enable them to use the earth's magnetic field for navigation.

One of the mysteries of the animal kingdom is the semiannual migration of birds, sometimes over thousands of miles, often over long stretches of water. For several years, I was very fortunate to have a pair of scarlet tanagers choose the woods behind my home for their summer residence. Scarlet tanagers winter in

South America, but this pair returned each year to my small lot in upstate New York. Other birds migrate over even larger distances. How do they find their way?

Migrating birds employ their navigational skills only twice a year, but homing pigeons have the year-round ability to return to their home lofts after being released many miles away. For selected and trained pigeons of the U.S. Signal Corps, flights of 1,000 miles were routine, and the record flight home was 2,300 miles. (In this respect, homing pigeons are much more impressive than Lassie, but in movies and television looks are often more important than talent.)

The sun, wind, stars, and other features of the natural environment are part of the navigational clues used by migrating or homing birds. But, starting in the 1960s, evidence began to mount that they may also use the earth's magnetic field. A group in Frankfurt, Germany, raised this possibility with experiments on European robins. Kept in circular test cages in laboratories offering no visual cues, the robins showed a preferential orientation toward certain directions, and these preferences could be changed by applying fields from electromagnets. But the results showed great variability, and many scientists remained skeptical. It was the experiments by William Keeton of Cornell University on homing pigeons that convinced ornithologists and animal behaviorists throughout the world that birds had a magnetic sense.

Keeton's classic paper was published in 1971. He glued small magnets to the backs of some pigeons and nonmagnetic (brass) bars of nearly the same size and weight to the backs of others. The birds were released many miles from their lofts and observed until they vanished from sight. On overcast days, when the sun could not be used for directional cues, most of the birds with nonmagnetic bars on their backs disappeared in the homeward direction, as did unburdened birds. But many of the birds with magnets on their backs did not. Keeton argued that the field from the magnets masked the earth's field, blocking the birds' ability to use its magnetic sense to determine orientation.

It was a striking experiment, with a striking result, and has been cited many times since as the most clearcut observation that birds have a magnetic sense. Later experiments by others even suggested that pigeons could detect not only the direction of the earth's field but also slight variations in the field and in this way get information on location. (To know the way home, it helps to have not only a "compass" but also a "map" that tells you where you are.) Inspired by Keeton's results, scientists have reported evidence of magnetic navigation by many other birds and animals, including turtles, dolphins, and whales.

Keeton began to have some doubts about his own results, however. Magnets apparently did not always confuse the pigeons, and in a 1972 paper he noted "the disturbing variability found in the results." He went on to test several hundred new birds in a great variety of situations. His new data, gathered from 1971 to 1979, showed no statistically significant effect of the magnets. These negative results, however, were not published until 1988, eight years after his death. (As in many fields, studies with negative results are less likely to be published than those with clearcut positive results.) Although the discrediting of one research paper should not discredit an entire field of study, Keeton's original paper was so influential in the study of avian navigation that these new results have thrown considerable doubt on the importance of magnetic cues to pigeons and other animals.

Honeybees have been almost as popular as pigeons in studies of the magnetic sense of animals. The orientation of the "waggle dance" that forager bees use on returning to the hive, believed to communicate to other bees the direction of a good food source, is said to be affected by the earth's magnetic field. Other studies suggest that the earth's field influences the orientation of the sheets of comb that bees build in the hive, and even that the bees' daily biological rhythms may be set in part by the slight daily variations in the earth's magnetic field (caused by the effects of solar radiation on the ionosphere).

A group in Honolulu has recently conducted experiments on honeybees somewhat analogous to those of Keeton's studies of pigeons. Bees were anesthetized by chilling, and tiny magnetic or nonmagnetic rods were glued to their abdomens with rubber cement. Bees conditioned by exposure to a magnetic field from a small electromagnet to select one food source over another were less able to select that food if they wore the tiny magnetic rods but not if they wore the copper rods. The researchers' conclusion, analogous to Keeton's for pigeons, was that the tiny magnets confused the magnetic sense of the honeybees. I personally found the statistics in these published results rather unconvincing, although I must admit that honeybees are more difficult experimental subjects than the inanimate metals and ceramics that I usually study, which are much more consistent in their behavior.

One professor at Manchester University in England believes that even humans have a sense of direction derived from the earth's magnetic field and devised Keeton-like experiments to demonstrate it. R. Robin Baker drove busloads of blindfolded students many miles from Manchester, using circuitous routes, and then tested whether they could point in which direction their dorms were located. Students with magnets on their heads reportedly did less well on this navigational test than students with similar nonmagnetic objects on their heads.

In his book *Human Navigation and Magnetoreception*, Baker wrote, "So many animals are now known to possess a magnetic sense that I confidently believe the final search will be to find an animal that is magnetically blind." Even scientists who firmly believe that birds and bees have a magnetic sense, however, are skeptical of Baker's results, which others have failed to replicate. My own judgment is that if a few Cornell pigeons could have misled Professor Keeton, a busload of Manchester students could have misled Professor Baker.

The most incontrovertible evidence for a magnetic sensor in a biological organism is the biocompass of several varieties of marine bacteria (Chapter 3). Here the physical function of the chain of magnetite particles is clear: the earth's magnetic field produces sufficient torque on the bacteria to rotate them into alignment. (The orientation of even dead bacteria can be controlled with a magnetic field.) Similar chains of magnetite have been found in some algae, in sufficient quantity to provide enough torque to align these much larger organisms.

Magnetite crystals have also been found in pigeons, honeybees, and even in the human brain, but in proportionately much smaller amounts than in the bacteria and algae. If these larger animals indeed have a magnetic sense of direction, the mechanism must be more indirect, presumably acting via the nervous system. Models have been proposed in which magnetic torques on the magnetite particles produced by the earth's field result in changed ion flow through the membranes of neurons, or changed electrical paths at synaptic junctions. Researchers in Taiwan have recently observed that the magnetite in the abdomens of honeybees is located in cells connected to the bees' nervous systems.

As I noted in regard to the effects of ELF fields on biological systems, the existence of a magnetic sense of direction does not depend on our ability to explain its mechanism. Despite the recent discrediting of Keeton's classic work on pigeons, other studies give strong evidence that many birds, fish, and mammals are capable of detecting magnetic fields of a gauss or more. For example, very recent research at the University of North Carolina has revealed that the orientation of hatchling loggerhead sea turtles, swimming in total darkness, can be influenced by magnetic fields from electromagnets. Marine biologists guess that this magnetic sense helps guide the hatchlings from their nests on the beaches of Florida and Caribbean to their feeding grounds in the mid-Atlantic. Perhaps. But how hatchling turtles and other members of the animal kingdom are able to detect magnetic fields, and how important that ability is to their survival, remain among the many mysteries of nature we have yet to solve.

///// SOURCE OF THE FORCE

16

///// **Clark Kent and Superman**

A major theme of the preceding chapters has been that magnets provide the driving force of much of today's technology. The magnetic force, in tandem with its alter ego, the electric force, has also played a central role in the development of modern physics.

The mysterious attraction between lodestones and iron was not the only example of "force at a distance" known to the ancients. They also knew that amber (fossilized tree resin), when rubbed, could attract feathers and other light objects. "Great has ever been the fame of the lodestone and of amber in the writings of the learned," wrote William Gilbert in 1600. He noted that "electric" forces (from *electron*, Greek for amber) were weaker than magnetic forces but were more general. Many materials other than amber could be rubbed to produce electric forces, and many things could be attracted by electric forces. Lodestones attracted only iron.

We know today that all matter consists of negatively charged electrons and positively charged nuclei and that usually these exist in balance so there is no net charge. But rubbing amber or other materials can transfer a few electrons and produce a small net charge, and the electric field from the charged material can attract light objects. As a way of differentiating this force from magnetic forces produced by moving charges, it is often called an *electrostatic* force. It is the "static cling" that holds clothes to the body, bemoaned in many TV commercials for laundry products, and it can draw a spark between your hand and the doorknob if you've been walking across a rug. But electrostatic forces are not all bad. In photocopiers, for example, they are responsible for the transfer of ink to paper.

In *De Magnete*, Gilbert clarified the many differences between electrostatic and magnetic forces. Electricity and magnetism were considered separate phenomena until 1820, when Oersted, Denmark's "other Hans Christian," discovered that an electric current produced a magnetic field (fact 6). Soon thereafter, Michael Faraday discovered that a changing magnetic field induced an electric current (fact 8). Electricity in motion produces magnetism, and magnetism in motion produces electricity.

The observations of Oersted, Ampère, and Faraday on the connections between electricity and magnetism led James Clerk Maxwell to his famous four equations, which are today proudly displayed on the T-shirts of many college science students. These equations indicated that waves of electromagnetic energy could travel through space with a velocity that equalled the velocity of light, which had already been measured. This led Maxwell to the remarkable conclusion, hitherto unsuspected, that light consisted of oscillating electric and magnetic fields. His conclusions unified electricity, magnetism, and optics and soon led Hertz to generate electromagnetic waves at radio frequencies. Today a wide range of phenomena are recognized as electromagnetic waves of different frequencies and wavelengths (Figure 8.2), all part of "Maxwell's rainbow." In 1964 Richard Feynman wrote, "From a long view of the history of mankind—seen from, say, ten thousand years from now—there can be little doubt that the most significant event of the 19th century will be judged as Maxwell's discovery of the laws of electrodynamics. The American Civil War will pale into provincial insignificance in comparison with this important scientific event of the same decade."

Although electricity and magnetism are closely related, most physics courses still make a clearcut distinction between electric fields and magnetic fields. Charges create electric fields even when stationary, and electric fields exert forces (electrostatic forces) even on stationary charges. In contrast, charges create magnetic fields only when moving, and magnetic fields exert forces only on moving charges. As these concepts are normally taught, electric fields and magnetic fields, although related, are distinct.

In *There Are No Electrons: Electronics for Earthlings,* Kenn Amdahl suggested that electric fields and magnetic fields are the Clark Kent and Superman of science: "like Lois Lane, we may never be able to exactly figure out what's going on, but we sure can tell there's a mysterious relationship between the two . . . It's almost like they were different costumes on the same character." To clarify why electric and magnetic fields are not distinct—why they are really the "same character" in a "different costume"— requires a careful consideration of relative motion, the topic that made Albert Einstein famous.

Einstein remained fascinated with magnetism long after his childhood wonder at the mysterious motion of a compass needle. In 1895, when he was only sixteen, he sent a five-page paper to his uncle on the topic of the magnetic fields produced by permanent magnets and by electric currents. And Faraday's law of magnetic induction, the "fact 8" I have been citing over and over again throughout this book, was influential to Einstein's development of the theory of relativity. This is clear from the title of the 1905 paper in which he introduced special relativity: "On the electrodynamics of moving bodies."

In the first paragraph of that momentous paper, Einstein discusses "the reciprocal electrodynamic action of a magnet and a conductor." If you consider the magnet stationary and the conductor moving, the induction of current in the conductor can be explained in terms of the force produced by the magnetic field of the magnet on the charge in the moving conductor. If you instead consider the conductor stationary and the magnet moving, the induced current is explained by the electric field produced by a changing magnetic field. Yet either way, of course, the same current is induced. Two observers, one holding the magnet and one holding the conductor, may interpret the same phenomenon differently (Figure 16.1).

Consider an even simpler case. A beam of electrons travels at constant speed past an observer, who, using a compass, detects a magnetic field produced by the moving charge. A second observer, traveling along with the electrons, sees only stationary charge and electric field—no magnetic field, no compass deflection. Two observers, moving with respect to each other, see different fields. Both see an electric field, but only one sees a magnetic field. Other examples can be devised in which one observer sees only an electric field and another sees only a magnetic field. Two different costumes, but the same character: an electromagnetic field. Clark Kent *is* Superman! (Like most analogies, this one has its limits. Superman and Clark Kent were seldom seen together, although electric and magnetic fields often are.)

Magnetic fields and forces are, from a fundamental point of view, simply relativistic effects of electric charge in motion. As it happened, an understanding of the relationship between electricity and magnetism led Einstein to the theory of relativity. But it could have happened differently.

One of the early American responses to Einstein's 1905 article introducing special relativity was an article of 1912 in the *American Journal of Science* written by Leigh Page, then a young instructor at Yale. Page showed that by applying the principle of special relativity to the law of electrostatic forces (Coulomb's law, developed in the late eighteenth century by Charles-Augustin de Cou-

NOW AGAIN... ONE MORE TIME... WE'LL DO THE FARADAY EXPERIMENT... BUT THIS TIME IN 𝕆𝕌𝕋𝔼ℝ 𝕊ℙ𝔸ℂ𝔼, SO WE CAN'T TELL WHO IS "REALLY" MOVING. WE KNOW ONLY THAT WE ARE MOVING 𝐑𝐄𝐋𝐀𝐓𝐈𝐕𝐄 TO EACH OTHER.

I THINK I AM STATIONARY, AND RINGO IS MOVING. I DETECT A MAGNETIC FIELD, BUT IT CAN'T MOVE THE CHARGES, SO THERE MUST BE AN ELECTRIC FIELD ALSO, CAUSED BY THE CHANGING MAGNETIC FIELD.

RINGO THINKS HE IS STATIONARY AND I AM MOVING. HE DETECTS ONLY A MAGNETIC FIELD AND MOVING CHARGES, WHICH ACCOUNT FOR THE INDUCED CURRENT.

RINGO AND I DISAGREE ON WHAT FIELDS ARE PRESENT!

Figure 16.1 The *Cartoon Guide to Physics* explains that different observers will detect different fields in a Faraday induction (fact 8) experiment. This observation implies that electric and magnetic fields are just different aspects of a single field, the electromagnetic field, and inspired Einstein to develop the special theory of relativity. (© 1990 by Lawrence Gonick and Arthur Huffman. Reprinted by permission of HarperCollins Publishers, Inc.)

lomb), he could directly derive the existence of magnetism! He could derive the magnetic fields produced by electric currents (fact 6), the force on moving charges and current-carrying wires produced by a magnetic field (facts 9 and 10), and the electric fields produced by changing magnetic fields (fact 8). In other words, all the discoveries of Oersted, Ampère, and Faraday could have been predicted simply from Coulomb's law of electrostatic forces and special relativity! Instead of using electricity and magnetism to deduce relativity, we could have used electricity and relativity to deduce magnetism. It didn't happen that way, but it could have.

Physicist Edward Purcell once was asked whether there were any engineering applications of relativity. He pointed out that there were many, since much of technology depends on magnetism, and magnetism is a result of relativity.

When physicists refer to the four basic forces of the universe, they mean gravitation, strong and weak nuclear forces, and the electromagnetic force. That used to bother me, since my college courses had included no relativity. Although I knew that electric and magnetic forces were closely related, they still seemed distinct to me. Why were they lumped into one? I learned later that physicists don't consider electric and magnetic forces as two different forces because relativity reveals their common identity. Physicists are more perceptive than Lois Lane.

During an interview on the BBC, Richard Feynman was asked why or how two magnets repelled each other. Feynman responded, "It depends whether you're a student of physics or an ordinary person who doesn't know anything. If you don't know anything at all, about all I can say is that there's a magnetic force that makes them repel . . . You're not at all disturbed by the fact that when you put your hand on the chair it pushes you back. But we found out by looking at it that that's the same force." (Translation: the force a chair exerts on your hand is the electric repulsion between electrons in the atoms of the chair and electrons in the atoms of your hand, and we know from relativity that electric and magnetic forces are two manifestations of the same basic force.)

///// Alice and the Red Queen

Einstein's theory of relativity was a severe blow to "common sense." Time, length, mass, and electric and magnetic fields were no longer absolute quantities; instead, they depended on the velocity of relative motion between the observer and the observed. From a practical point of view, the effects of relativity could usually be ignored by engineers as long as they were concerned with systems in which relative velocities remained far less than the speed of light. But the concept that time itself was not absolute was particularly unsettling. And we barely had absorbed some of the strange results of relativity before the next challenge to "common sense" made its appearance: *quantum mechanics.*

When I first introduce quantum mechanics to sophomores majoring in materials science and engineering, I usually start with the passage in Lewis Carroll's *Through the Looking Glass* where the Red Queen tells Alice her age:

> "I'm just one hundred and one, five months and a day."
>
> "I can't believe *that!*" said Alice.
>
> "Can't you?" the Queen said in a pitying tone. "Try again: draw a long breath, and shut your eyes."
>
> Alice laughed. "There's no use trying," she said: "one *can't* believe impossible things."
>
> "I daresay you haven't had much practice," said the Queen. "When I was your age, I always did it for half-an-hour a day. Why, sometimes I've believed as many as six impossible things before breakfast."

I then tell my students that the Red Queen would have been an ideal student of quantum mechanics. Some aspects of quantum mechanics certainly *seem* impossible, because they are contrary to the "common sense" we've developed observing objects near our own size level. At the atomic level, at dimensions less than a nanometer, objects behave differently. Quantum mechanics was developed by theoretical physicists but it triumphed, as James Gleick has written, "not because a few theorists found it mathematically convincing, but because hundreds of materials scientists found that it worked."

Physicists were familiar with the concept of waves, including water waves, sound waves, and light waves, and had learned from Maxwell that what was "waving" in light were electric and magnetic fields. Yet new experiments showed that light, when beamed onto the surface of a metal, delivered energy to the electrons of the metal only in discrete, or *quantized*, amounts, suggesting that light was composed of particles, soon called *photons*. (Einstein received the Nobel Prize for explaining that, not for relativity!) Light sometimes acted like a wave, sometimes like a particle.

How about electrons, the basic units of negative electricity? Electrons had originally been understood to behave like beams of particles, but in new experiments electron beams were bounced off metal crystals and formed regular patterns ("diffraction patterns"), which indicated that electrons were acting like waves. Both light and matter had been found to exhibit "wave-particle duality." Also baffling was the growing body of data on the optical properties of atoms. Each element of the periodic table absorbed and emitted light in a characteristic series of frequencies. Classical physics was unable to explain the structure, or even the existence, of atoms, much less their complex optical properties.

Several mathematical approaches were developed to explain the strange properties of matter at the atomic scale, the most successful being the wave equation of Erwin Schrödinger. Leaving his wife at home in Zurich, Schrödinger took his 1925 Christmas vacation in the Swiss Alps, bringing along an "old girl-friend" for inspiration. The mystery woman remains anonymous, but she must have been very inspiring, because Schrödinger returned home with a beautiful nonrelativistic equation for electron waves that successfully explained the optical properties of the hydrogen atom. Three weeks later, he submitted his results for publication. To quote Schrödinger's biographer, "This paper has been universally recognized as one of the greatest achievements of twentieth-century physics."

But Schrödinger had not been successful in his attempts to incorporate relativity into his treatment. This advance was made in 1928 by Paul Dirac, whose theory of the relativistic quantum mechanics of the electron showed that the electron must have a spin (an idea that had been suggested earlier by experiments). Dirac's theory also yielded a value for the magnetic strength of an electron that was in fair agreement with experiment. Thus the electron spin, source of the magnetism in young Einstein's compass needle, had been shown to be a natural consequence of Einstein's theory of relativity and the quantum-mechanical nature of matter.

Dirac's paper also had an important spin-off. Some solutions to his equations corresponded to electrons of negative energy but could also be interpreted as implying the existence of an anti-electron, a particle with the same mass as an electron but with opposite (positive) charge. The anti-electron, called a *positron*, was discovered in 1932. The discovery of anti-protons and other antimatter particles followed. When matter and antimatter meet, they disappear with the release of energy; when electrons and positrons collide, for example, energy is released in the form of gamma-ray photons, as described in Chapter 14's discussion of PET. The reverse is also possible: matter and antimatter may simultaneously be created by the annihilation of high-energy photons (Figure 16.2). Whereas the former process converts mass into pure energy, the latter converts pure energy into mass.

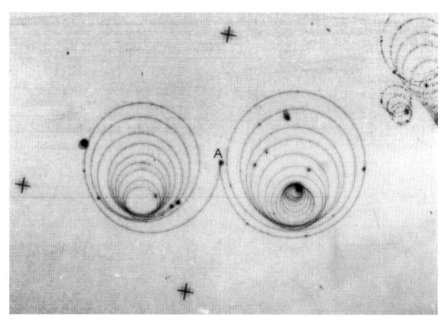

Figure 16.2 Tracks revealing the creation of a positron (anti-matter!) and an electron at point A from a gamma-ray particle (which entered from above, but left no track because it is uncharged). A magnetic field perpendicular to this view produces circular motion of both particles (see fact 9 and Figure 5.1B), but they curve in opposite directions because of their opposite charge. As they gradually slow down, they move in smaller and smaller circles.

Dirac's successful prediction of the positron lends credence to his 1931 proposal of the existence of *magnetic monopoles*. Positive and negative electric charges can exist in isolation, why not north and south magnetic poles? But extensive experimental searches for evidence of magnetic monopoles, by many investigators, have so far been unsuccessful. Some theorists believe that monopoles existed for a very short time in the early history of the universe and may some day be produced in particle accelerators capable of much higher energies than are available today. Perhaps. But for now, we can continue to believe what we were taught in school: magnetic poles always occur as north-south pairs.

///// Dave Barry and Virtual Effluvium

Columnist Dave Barry recently tested the "Science IQ" of his readers with a few multiple-choice questions, one of which was:

> What is magnetism?
> (a) Invisible rays that shoot out of a compass.
> (b) The force that causes dogs to bark when you ring the doorbell.
> (c) The molecular attraction that forms between refrigerators and little ceramic vegetables.

Barry wrote that the correct scientific answer is "d. No opinion," but I believe he was kidding, as is his wont. Answers (b) and (c) have some merit, but Barry's "invisible rays" in answer (a) are consistent both with the ideas of ancient times and with the modern ideas of Feynman and others as incorporated in the theory of quantum electrodynamics (QED).

We saw earlier that, from the reductionist viewpoint of theoretical physics, the electromagnetic force between two magnets is really the same electromagnetic force that you feel when your hand touches a chair. But it sure *looks* different! What amazes us about the forces between magnets is that they act at a distance (fact 4), *without touch.*

To explain the magical "force at a distance" between lodestones and iron, Epicurus (341–270 B.C.) blamed "atoms and indivisible bodies that flow from stone and from iron." Even though Epicurus couldn't see anything flowing between the lodestone and the iron, he felt that something must be delivering the force! Other early writers proposed that "magnetic effluvium" or "heavy exhalations" flowed across the distance separating two magnets, but Gilbert argued in 1600 that if effluvium flowed between magnets, it must be "incorporeal," because "though solid and very dense bodies, or blocks of marble, stand between, they do not hinder the passage of the potency" (fact 4 again).

According to modern QED theory, electromagnetic forces between charged particles are delivered by the exchange of "virtual photons," quanta of electromagnetic energy that, like ordinary photons, travel at the speed of light but, unlike ordinary photons, cannot be directly detected. They're "incorporeal" and go right through Gilbert's "blocks of marble." Described in this way, they don't seem much more sophisticated than the effluvium of Epicurus. But behind them stands a highly mathematical, very quantitative theory that can be compared with experimental measurements. And the agreement is excellent.

QED theory seems strange even to its practitioners, as evidenced by the title of Feynman's popularized treatment of the subject, *QED: The Strange Theory of Light and Matter.* But it works. Dirac's original treatment of the relativistic quantum mechanics of an electron gave a value for the magnetic strength of an electron of exactly 1 in certain units. In the same units, the best experimental value for the magnetic strength of an electron is 1.0011596522. Dirac's original theory, the starting point for QED, did not take into account all the effects of the electron's interactions with light. Modern QED, including these effects, gives a theoretical value of 1.0011596525. The disagreement with experiment is less than one part in a billion. Other electromagnetic properties have been measured with comparable accuracy, and QED calculations agree with the measurements. As Feynman says, "These numbers are meant to intimidate you into believing that the theory is probably not too far off!"

The initials QED for quantum electrodynamics seem especially appropriate to scientists because they are also used in the proofs many of us remember from our early studies in mathematics. At least in my high school days, the geometry course consisted largely of proving various equalities and geometric relationships using the axioms of Euclid and other relationships we had previously proven. Each homework or exam problem would start with a statement of the relationship to be proven, which would be followed by a series of logical steps that led from axioms or previously proven relationships to the desired conclusion. To show that we had proven what we had intended to prove, we then wrote Q.E.D., for *quod erat demonstrandum*, Latin for "which was to be demonstrated." So QED seems just right for a highly successful scientific theory.

Most college courses on electricity and magnetism never even mention QED and virtual photons. They describe electric and magnetic forces in terms of invisible electric and magnetic fields, as I've been doing throughout this book. ("Force fields" are also a staple of TV science fiction.) Engineers design motors and television sets and materials scientists develop improved magnets and superconductors without using QED theories. In fact, most macroscopic phenomena involve so many electrons and photons that QED wouldn't be much direct help. But for looking, very accurately, at phenomena on the atomic scale that involve only a few particles, QED works.

The father of QED, Paul Dirac, wrote in 1929, "The underlying physical laws necessary for the mathematical theory of a large part of physics and the whole of chemistry are thus completely known, and the difficulty is only that the exact application of these laws leads to equations much too complicated to be soluble." But that's a large difficulty. From an engineering viewpoint, a theory that can't be used to solve real problems, even though it may be correct in principle, is of little value. But many macroscopic properties depend on the electromagnetic interaction of particles at the atomic level, which can be treated very accurately by QED. QED has advanced considerably since 1929, and electronic computers now allow much more complex calculations than could be imagined in 1929. The concepts of electric and magnetic fields, and their behavior as described in Maxwell's equations, remain adequate to solve most of our engineering problems. But it's comforting to know that these equations are firmly based on a highly successful microscopic theory, particularly as some parts of modern engineering focus on smaller and smaller dimensions.

Virtual photons, the "messenger particles" of electromagnetic fields, are a weird concept. QED calculations involve other weird concepts, including the appearance and disappearance of virtual electron-positron pairs and particles that travel backward through time. (Even the Red Queen would have difficulty with QED.) Is all this "real"? Are invisible electric and magnetic fields "real"? Whether we like them or not, as descriptions of nature they work. In his book describing QED, Feynman warned his readers,

> I'm going to describe to you how Nature is—and if you don't like it, that's going to get in the way of your understanding it. It's a problem that physicists have learned to deal with: they've learned to realize that whether they like a theory or they don't like a theory is *not* the essential question. Rather, it is whether or not the theory gives predictions that agree with experiment. It is not a question of whether a theory is philosophically delightful, or easy to understand, or perfectly reasonable from the point of view of common sense. The theory of quantum electrodynamics describes Nature as absurd from the point of view of common sense. And it agrees fully with experiment. So I hope you can accept Nature as She is—absurd.

The quantitative success of QED in describing electromagnetic forces (and therefore, in principle, "the whole of chemistry") inspired physicists to develop parallel theories for the rest of physics—nuclear and gravitational forces. Virtual photons were the messenger particles of the electromagnetic field, delivering "force at a distance" by traveling between two interacting charged particles. Building on this concept, physicists invented other messenger particles to deliver the other forces: massive W and Z particles to deliver the weak nuclear force, massless "gluons" to deliver the strong nuclear force that binds quarks together to form protons and neutrons (a theory called QCD, or quantum chromodynamics), and "gravitons" to deliver the gravitational force.

The theory of the weak nuclear force was developed along

lines similar to the theory of the electromagnetic force (QED), with the two forces "unified" into an "electroweak" force at very high energies (energies corresponding to our universe less than a billionth of a second after the Big Bang, but energies accessible in high-energy particle accelerators). Theories attempting to unify the electroweak force with QCD are called Grand Unified Theories, or GUTs, but they have the disadvantage that the energies at which these forces are believed to unify are well beyond experimental capabilities. Yet such theories, and others that attempt to incorporate the comparatively weak gravitational force ("Theories of Everything"), relate not only to models of the elementary structure of matter but also to the formation of all matter of the universe in the Big Bang. Thus both the particle theorists who "think small" and the cosmologists who "think big" have a common interest in the unification of the basic forces of the universe at higher energies.

You'll have to read other books to learn about relativity, quantum mechanics, QED, QCD, particle physics, and cosmology. This book is about magnets. But all these challenging areas of modern science are closely related to the mysterious "force at a distance" between lodestones and iron and "between refrigerators and little ceramic vegetables." The mysterious magnetic force that drives our motors and loudspeakers ("movers and shakers") and many other components of today's technology is also at the heart of our search for basic truths about the formation and structure of the universe and all that is in it, including ourselves. Everywhere we look, The Force is with us.

Quod erat demonstrandum. Q.E.D.

///// PULLING IT TOGETHER

17

///// Magical Mystery Tour

We've covered a lot of ground together in our tour through the magical world of magnets. (I am now assuming that you're not one of those readers who looks at the last chapter first, "to see how it comes out.") We started with a few familiar aspects of magnets and the magnetic force (compasses, toys, paper clips, and refrigerators), a brief introduction to the many hidden magnets in a familiar example of modern technology (the VCR), and ten basic "facts about the force." We discussed natural magnets (lodestones, other magnetic rocks, and the earth itself) in Chapters 2 and 3 and manmade magnets (permanent magnets, electromagnets, and superconducting magnets) in Chapters 4 and 5. And in Chapter 6 we explored some of the physics and materials science of magnetic and superconducting materials.

Next we skimmed through some of the many applications of magnets and magnetic forces—motors and speakers (Chapter 7), transformers and TVs (8), recording (9), levitation (10), and warfare (11). As "tour guide" for this survey, I surmised that five chapters on technology may be enough for some readers, so I then steered the discussion to magnets at play (12) and mesmerism (13) and next covered magnets in medicine (14) and biology (15). I tried in Chapter 16 to complete the "big picture" with a review of the important place the magnetic force holds in major developments of twentieth-century physics, including relativity, quantum mechanics, and QED. In these final few pages, I'd like to pull together a few general observations from the myriad sights that we've seen on our tour.

1. Magnets have a long and colorful history. From the days of Plato and Pliny and the Chinese magicians to the days of Dave Barry, Dick Tracy, and *Star Trek*, people have been fascinated by the mystery of magnetic forces. Magnets have stimulated the imaginations of writers (Ibn Hazm, Ben Jonson, Jonathan Swift, and William Gilbert) and scientists (Faraday, Einstein, Feynman, and an earlier William Gilbert). Magnets have been used by charlatans to mislead the gullible (the "Celestial Bed" of James Graham, the psychokinesis of Uri Geller, and the supermagnets that today are claimed to cure cancer, sharpen razor blades, and increase gas mileage), and they have been used by explorers to discover strange new worlds (Christopher Columbus, Anton Mesmer, and today's scientists who use MRI and MEG to study the inner workings of the human brain).

Of course what makes the history of magnets so colorful is not the magnets but the colorful characters. On our whirlwind tour, we haven't spent enough time with any one character to fully appreciate his or her life story, but in the notes on sources following this chapter I'll refer you to some intriguing biographies. One of the most fascinating figures is Mesmer, about whom much has been written. Although his cures are today viewed as some combination of hypnosis and faith healing, his biographers agree that Mesmer sincerely believed in his theories of "animal magnetism." Many well-known figures enter into his life story, including Mozart, Jefferson, Franklin, Queen Marie Antoinette, and Madame Du Barry. The setting of his major triumphs and defeats was Paris on the eve of the French Revolution. His ideas led, among other things, to the development of hypnotherapy, Christian Science, and chiropractic. For all these reasons, Mesmer's story is one of the most interesting in the history of magnetism.

2. There are magnetic fields much higher than, and much lower than, the earth's field. The earth's field, you'll recall, is about half a gauss. Scientists and engineers have been able to produce fields much higher than that and measure fields much lower than that. The saturation magnetization of iron is about 20 kilogauss, and some iron-core electromagnets can reach roughly twice that field. Niobium-titanium superconducting magnets can reach 100 kilogauss, and niobium-tin magnets 200 kilogauss. Hybrid magnets consisting of a Bitter-style central insert and a superconducting "outsert" have produced steady magnetic fields of about 400 kilogauss, approximately a million times stronger than the earth's field. Pulsed magnets have reached a megagauss, and implosion techniques using explosives to compress magnetic fields have reached 10 megagauss—but for less than a microsecond. That's beginning to approach truly "astronomical" levels: the fields observed in white dwarf stars can reach 100 megagauss, and neutron stars have fields of a teragauss (a mega-megagauss).

The magnetic fields obtainable from ferromagnetic materials are limited by their saturation magnetization. When all the atomic magnets of these materials are lined up, they've done all they can do. Copper-wound electromagnets are limited by electric power consumption and the need for cooling to counteract the resistive heating. Superconducting magnets are limited by their critical fields and critical currents, and all high-field electromagnets are limited by the ability of the windings to withstand the large electromagnetic forces. But the new U.S. National High Field Magnet Laboratory and other high-field laboratories in Europe and Japan continue to move toward higher fields. The goal is not to appear in the Guinness Book of Records but to allow scientists to study the properties of matter under extreme conditions, enabling them to "boldly go where no one has gone before."

For scientists studying fields much less than the earth's field, the problem is not to create such fields but to measure them in the presence of other, much larger fields. In the laboratory, and even in your home with its many appliances, power-frequency magnetic fields may reach a few milligauss, and variations in the earth's magnetic field can also reach that level. The magnetic field from your beating heart, measured outside of your body, is about a microgauss, a bit less than the average level of magnetic fields in interstellar space. The magnetic fields from brain activity are much weaker, roughly a nanogauss, but they can be measured with SQUIDs even though they are a million times weaker than the magnetic noise that surrounds us. Modern-day SQUIDs, attached to gradiometer circuits to screen out noise, are much more sensitive than the MADs developed in World War II to detect submarines. (SQUIDs are MADDer than MADs.)

These high and low magnetic fields all appear on the logarithmic scale at the end of Chapter 4, where you can also look up the nanos, micros, megas, and teras if you've forgotten how to interpret those useful prefixes.

3. *Magnets have widespread importance in modern technology.* Decorative and promotional magnets are a large and growing business, and many children's toys and "executive toys" for adults employ magnets. But much more important to our daily lives are the thousands of hidden magnets that provide the driving force of today's technology.

Forces from magnets drive the many motors and speakers (movers and shakers) that we use every day to convert electrical energy into motion and sound in our homes, cars, and offices. Forces from magnets give us the images on our TV and computer screens, and they provide physicians with images of the inner

THE LIST IS ENDLESS

In addition to the hundreds of technological applications of magnets mentioned in this book, there are hundreds more that could be added. A few examples:

- vending machine magnets that identify coins by induced eddy currents
- compact screw-in fluorescent induction lamps (powered with ferrite antennas)
- ballasts for normal fluorescent tube lights
- video cameras
- magnetic microphones
- roadway car sensors that operate traffic lights
- industrial ultrasonic cleaners
- impact printers
- racetrack starting gates
- magnetic stirrers and other coupling devices that transmit rotational motion across barriers
- induction furnaces that heat via eddy currents
- AC levitation melting (containerless melting of highly reactive metals, in which AC fields provide both heating and levitation)
- laboratory measuring instruments
- eddy-current crack detectors
- magnetic fluids (magnetic particles dispersed in various fluids, called ferrofluids) for crack detection
- seals for rotating shafts
- magnetic inks
- hotel room keys
- metal detectors
- finding a needle in a haystack

structure of our bodies and high-energy physicists with images of the inner structure of matter. Forces from magnets levitate and propel watt-hour meters and high-speed trains, heat our morning coffee in microwave ovens, and pull sounds and images from the air in our radio and TV receivers. Forces from magnets write and read information in audio and videotapes, hard and floppy disks, credit and ATM cards. They also play an increasingly important part in modern warfare. And forces from magnets drive currents in generators and transformers to deliver electricity to our homes.

The ubiquity of the magnetic force in technology is not too surprising from the viewpoint of fundamental physics, which tells us that there are only four forces in the universe. The strong and weak nuclear forces are used sparingly in technology, and although we certainly use the gravitational force from the earth, we can do little to vary it, and gravitational forces between small objects are extremely weak. That leaves only the electromagnetic force in its two guises—electrostatic and magnetic. The electrostatic force is used primarily in force by touch, including the impact of burning and expanding gases that push the pistons in your automobile engine. (The physicist recognizes force by touch as the electrostatic repulsive force between electrons.) But for controllable force without touch, it's hard to beat the magnetic force. It is therefore the magnetic force that drives much of modern technology. Wherever we go and whatever we do, The Force is with us.

4. Magnets are better than ever. Many scientists and engineers who work with magnets feel that magnets, like Rodney Dangerfield, "don't get no respect." Magnets, after all, have been around since ancient times and used in technology for over a century. To most of the general public, who see magnets only in toys and on refrigerators, magnets lack the glamor of recently discovered wonders like lasers, transistors, and superconductors. But "neo" (neodymium-iron-boron) permanent magnets, discovered only in 1983, are more than a hundred times more powerful than the carbon steels of the nineteenth century. As a result, hundreds of technological devices, especially motors and speakers, have been decreased in size and weight and increased in efficiency. Many devices that are practical today were impossible or impractical before the development of rare-earth magnets. Improvements in soft magnetic materials have also been impressive, having saved billions of dollars and many barrels of imported oil, through decreased energy losses. (Improved hard magnetic materials have also contributed to energy conservation, by allowing the replacement of electromagnets by permanent magnets in many motors.) Improvements in magnetic recording materials, both hard and soft, have been especially dramatic; bit densities in computer disks have increased by a factor of 100,000 in less than forty years. Magnets have indeed been with us for a very long time but, thanks to the efforts of materials scientists, they are much better and much more important to us than ever before.

The problem with public perception stems partly from the fact that progress in hard and soft magnetic materials has been slow and steady rather than sudden and dramatic. Another problem may be that we all get exposed to magnets early in school: since "familiarity breeds contempt," perhaps early familiarity breeds lifelong contempt. Finally, most magnets in modern devices are hidden, which means that few people are even aware of their importance. Out of sight, out of mind. One of my goals in writing this book was to increase public awareness of the importance of magnetic materials and the exciting recent advances in their properties.

Superconductors don't seem to have the same image problem. In the early days of high-temperature superconductors, they rated magazine cover stories and front-page articles in *The New York Times*. (Having "super" as part of their name probably didn't hurt, which is why some workers with "neo" magnets bill them as "supermagnets.") In our materials courses at MIT, superconductors are considered as magnetic materials, because their two major applications are to produce very high magnetic fields (as in MRI magnets) and to detect very low ones (as in SQUIDs). The development of high-field superconductors in the 1960s led to powerful superconducting magnets that first had an important impact on the research community, particularly in high-energy physics. The most important application of such magnets to the general public today is in MRI. The high-temperature superconductors may someday become important to technology, but their applications to date have been very limited.

In my classes I often use the demonstration of a rare-earth "neo" magnet levitating above a high-temperature superconductor. I note that the magnet was discovered in 1983, the superconductor in 1987. The superconductor received intense worldwide press coverage, and over 40,000 research papers have been written about it. The magnet was barely mentioned in the press, and research activity has generated less than 4,000 papers. Yet the high-temperature superconductor, darling of the media and research physicists, has few applications today, and it is unlikely ever to become even remotely as important technologically as the "neo" magnet, which already has hundreds, if not thousands, of important applications. (One colleague has termed the introduction of rare-earth magnets into modern technology "the silent revolution.") Permanent magnets are super-useful, but don't get no respect. Go figure.

5. It's a material world. Although our tour has focused on magnetic materials, it has also provided us a glimpse into the larger world of materials science. As important as magnets are, they are not the whole show. Recent advances in other materials classes (semiconductors for electronic and optoelectronic applications, glass fibers for optical communications, high-tech ceramics, polymers, and composite materials) have also had a major impact on modern technology. As Madonna reminded us in her trademark song, we live in a material world. Technological devices are limited by the properties of available materials, and advances in engineering materials have provided a driving force for advances in technology. As a result, the Federal Coordinating Council on Science, Engineering, and Technology, an advisory group with representatives from twelve Cabinet agencies and all major independent government agencies (NSF, NASA, CIA, etc.), has identified materials science as a national priority area for government support (along with global change research, high-performance computing, biotechnology, math and science education, and advanced manufacturing).

Materials science (see chapter 4) is a multidisciplinary applied science, closely coupled with engineering. Its subject is engineering solids—metals, semiconductors, ceramics, polymers, and composites of these material types. According to a display outside the headquarters of MIT's Department of Materials Science and Engineering, the four components of our field are PRocessing, PRinciples, PRoperties, and PRoducts. (The display was clearly created by an expert in PR.) Materials scientists use the PRinciples of chemistry and physics to understand how the PRoperties of engineering solids depend on the PRocessing methods used to manufacture them. And the ultimate goal is a very PRractical one: a PRoduct that can be used in a technological device.

The link between processing and properties is the *microstructure* of the material—how the atoms are arranged at the microscopic level. As we discussed in Chapter 6, the engineering properties of hard and soft magnetic materials—coercivity, energy product, permeability, AC energy losses—all depend on details of the microstructure, including crystal size and orientation and chemical and structural inhomogeneities. The same is true of critical currents of high-field superconductors and many other properties of engineering importance. The materials scientist therefore attempts to understand both the relation between microstructure and properties and the relation between processing and microstructure. A major component of modern materials science has been the development of new processing techniques, including rapid solidification (Chapter 8) and thin-film deposition (Chapter 9).

Materials science bridges the gap between the basic physics and chemistry of solids and their technological applications. When I worked on rare-earth permanent magnets at General Electric, I had to learn some of the quantum physics of magnetism in rare-earth elements, the chemistry of interactions between rare-earth elements and iron and cobalt, the metallurgical microstructures resulting from various processing treatments, the effect of these microstructures on magnetic domain-wall motion and magnetic properties, and the relation of these properties to the performance of rare-earth permanent magnets in motors, speakers, and other devices. When I worked on amorphous magnetic alloys, superconductors, alloys for jet engines, and other projects, I similarly had to learn the physics, chemistry, metallurgy, and engineering related to these specific materials and applications.

The head of my MIT department may be mad at me for telling you this, but you don't need a degree in materials science to do materials science. I worked in materials science for many years at General Electric, and I teach materials science at MIT, but my degrees were in applied physics. Many physicists and chemists interested in engineering solids, and many engineers interested in materials, work in materials science.

Although physicists, chemists, and engineers can all do materials science, those specifically trained in a materials science and engineering department have already been exposed to the foundations of the field—the relations between processing, microstructure, properties, and applications of engineering solids. And since both government and industry recognize the importance of this field to future technologies and international competitiveness, a degree in materials science and engineering is a degree with a solid future. (*That* should make my department head a bit happier with me!)

Modern processing techniques have begun to allow control of microstructures down to dimensions of nanometers (nanotechnology), and modern microscopes can study microstructures down to atomic dimensions (a few tenths of a nanometer). Whether done by physicists, chemists, engineers, or those specifically trained in the field, materials science has become more and more sophisticated. As a result, magnets and other engineering materials, and the technologies that depend on them, are better than ever. In at least this area of human endeavor, progress has been real and undeniable.

///// Wonders of the World

"The world is full of wonder to a young child" was the first sentence of Chapter 1. Young Einstein wondered at the motion of a compass needle, a motion that involved three separate mysteries that we've touched upon in this book: the magnetism of the compass needle, the magnetism of the earth, and force without touch. He wrote, "something deeply hidden has to be behind things."

The magnetism of the compass needle derives from the remarkable properties of iron. Of the hundred-odd elements in the periodic table, iron has the most stable nucleus, and it is the most common metal in the universe and in our planet. Each iron atom is a tiny magnet because of an unbalance in the electron spins, and the quantum-mechanical exchange force between net spins of neighboring atoms makes iron a ferromagnet, one of only three elements that are ferromagnetic at room temperature. The compass needle is not pure iron but steel, which contains carbon and perhaps some other elements. The resulting microstructure of the steel, visible only under a high-power microscope, obstructs the motion of magnetic domain walls and gives the steel enough coercivity to become a weak permanent magnet and maintain a fixed polarity. Thus the north pole of the compass needle remains a north pole, and that is why it was able to help Columbus, and other explorers, to find their way. The ferromagnetism of iron is certainly one of the wonders of the world.

The magnetism of the earth comes from electric currents flowing in molten metal 2,000 miles beneath our feet. The interplay between electricity and magnetism via facts 6 and 8 allows the earth's core to become a "self-excited dynamo," a mechanism also believed to produce the magnetic fields of other planets and the sun. The history of the earth's magnetic field, including many reversals, has been recorded in rocks around the globe that were formed millions and millions of years ago. Reading these rocks, and patterns of magnetic anomalies produced by magnetic rocks, has revealed the slow drift of continents and the spreading of the ocean floor. Bacteria and algae, and possibly many other creatures, have evolved with internal biocompasses that help them navigate successfully in their dangerous environments. And without the earth's field, a compass wouldn't have been much help to Columbus. Another wonder.

Force without touch mystified the ancients, who imagined "magnetic effluvium" flowing between lodestones and iron to deliver the force. The theory of gravitation, another force without touch, led to the concept of an invisible gravitational "field," and most electromagnetic calculations today are based on an invisible electric field derived from, and exerting forces on, electric charges and on an invisible magnetic field derived from, and exerting forces on, *moving* electric charges. Careful consideration of relative motion, however, revealed that electric and magnetic fields are, fundamentally, two different aspects of one field—the electromagnetic field. Merging of Maxwell's equations for electric and magnetic fields, Einstein's relativity, and quantum mechanics led to quantum electrodynamics—QED. In QED, the "magnetic effluvium" of the ancients reappears as "virtual photons" that serve as "messenger particles" to deliver the electromagnetic force. "Virtual photons" may sound phony, but QED allows very precise calculations of basic quantities, such as the magnetic strength of a single electron, that agree with experimental measurements to better than one part per billion. Without virtual photons to deliver the magnetic force, the magnetism of the earth's core would not have been able to move Einstein's compass needle. Despite our ability to calculate it accurately, force without touch remains a wonder.

Our ability to calculate, with great accuracy, the basic properties of nature is related to a fourth mystery in Einstein's childhood encounter with the compass. That mystery was Einstein himself, who as a child wondered about "something deeply hidden" and kept that wonder alive as an adult. His curiosity provided the driving force that moved him along the road that led, eventually, to quantum physics and relativity. The human mind is a complex electromagnetic instrument capable of writing novels and symphonies, conceiving scientific models of matter and the universe, and developing improved materials and devices. Advances in magnetic materials have given us new and powerful tools, including MRI, MEG, and PET, with which we are now studying the human brain. But we have barely begun to understand *that* mystery, one of the greatest wonders of the world.

The world is full of wonder, but not only to a young child. Physicist I. I. Rabi once said, "I think that physicists are the Peter Pans of the human race. They never grow up and they keep their curiosity." I'm no longer a young child, and I don't feel much like Peter Pan. But the world is still full of wonder:

- the ferromagnetism of iron, the most abundant metal in the universe;

- the magnetism of the earth, produced by currents in its liquid core;
- force without touch, delivered by virtual photons;
- and the intricate workings of the human mind.

Quod erat demonstrandum. Q.E.D.

///// SOURCES AND SUGGESTED READINGS

///// 1. A Magical Force

Einstein's brief autobiography, both in German and in English translation, appears in *Albert Einstein: Philosopher-Scientist* (New York: Tudor,
1949), edited by Paul A. Schilpp. Marlou Freeman's magnet collection
was described in the September 23, 1991, issue of *People* and in *The New
York Times* of July 25 and 28, 1991. *The Spy Who Loved Me*, starring Roger
Moore as James Bond and featuring Richard Kiel as Jaws, was released
by Metro-Goldwyn-Mayer/United Artists in 1977 and is available on
videotape. David Macaulay's *The Way Things Work* (Boston: Houghton
Mifflin, 1988) contains many well-illustrated explanations of video recorders, television picture tubes, motors, speakers, and other technological devices. It's also very funny.

For those who want to read further into the science of magnetism,
E. W. Lee, *Magnetism: An Introductory Survey* (New York: Dover, 1970;
originally published in 1963 by Penguin Books) is recommended. Although it was written too early to include superconducting magnets, it
is very readable and uses only a few mathematical equations.

///// 2. Romancing the Stones

Plato's "Ion" (Ion of Ephesus was a reciter of the epic poems of Homer)
appears in most collections of his dialogues, including *Great Dialogues*

of Plato (New York: Mentor, 1956), translated by W. H. D. Rouse and edited by E. H. Warmington and P. G. Rouse. Later in the dialogues, Socrates further clarifies his analogy: God runs his power through Homer, then through Ion, and thence to the audience, like the magnet's power runs through the series of iron rings. *Natural History,* an encyclopedic work covering all known aspects of natural science, was completed in 77 A.D. by Roman scholar Gaius Plinius Secundus (23–79), known as "Pliny the Elder." The quotation on magnets is in section XXV of book XXXVI, which covers minerals and precious stones. The writings of Ibn Hazm are discussed in detail by Anwar G. Chejne in *Muslim Spain, Its History and Culture* (Minneapolis: University of Minnesota Press, 1974). His references to magnets appear in *Ring of the Dove,* translated by A. J. Arberry (New York: AMS Press, 1953, reprinted 1981). Ben Jonson's play, *The Magnetic Lady,* dates from about 1630 and appears in most collections of his works.

The fascinating story of magnets and compasses in ancient China is told in volume 4, part 1, of Joseph Needham's *Science and Civilisation in China* (Cambridge: Cambridge University Press, 1962). "The Lodestone: A Survey of the History and the Physics" by M. Blackman (*Contemporary Physics,* vol. 24, no. 4, 1983, pp. 319–331) is recommended for a description of scientific measurements on lodestones. The microstructure of lodestones is described by P. J. Wasilewski in *Journal of Applied Physics,* vol. 50, no. 3, March 1979, pp. 2428–2430.

Addison's essay on the sympathetic telegraph appeared in the December 6, 1711, issue of *The Spectator* and was reproduced as Essay No. 241 in volume 10 of *British Essayists* (New York: Sargeant and Ward, 1810). The sympathetic telegraph and Strada's poem about it are discussed in F. V. Hunt, *Electroacoustics* (Cambridge, Mass.: Harvard University Press, 1954).

Magnetic prospecting is discussed briefly by H. Jensen and E. F. Peterson in the January 1948 issue of *Scientific American* and more extensively in P. V. Sharma, *Geophysical Methods in Geology* (Amsterdam: Elsevier, 1986) and in M. B. Dobrin, *Introduction to Geophysical Prospecting* (New York: McGraw-Hill, 1988). Magnetic anomaly data, based on aeromagnetic, marine, satellite, and ground measurements, are reviewed periodically in *Reviews of Geophysics,* including volume 29 (1991).

William Gilbert's *De Magnete* (1600) is available in a paperback repub-
lication (New York: Dover, 1991) of an earlier edition in English (New
York: Wiley, 1893), translated by P. F. Mottelay, that includes a biographi-
cal introduction and Gilbert's original illustrations. A brief biography
of Gilbert by J. Butterfield appears in volume 338 of *The Lancet* (December 21/28, 1991). The work of Gilbert, and of others who contributed to
the development of the science of magnetism, is described in G. L.
Verschuur, *Hidden Attraction: The Mystery and History of Magnetism* (New
York: Oxford, 1993). (Verschuur is an astronomer, and his book also includes a chapter on magnetic fields in space.) P. Benjamin, *A History of
Electricity* (New York: Arno Press, 1975, reprint of an 1898 edition by
Wiley) is outdated, but it gives an entertaining account of the history
of magnetism and electricity "from antiquity to the days of Benjamin
Franklin."

The effect of magnetic declination on Columbus's 1492 landfall is told
by Samuel Eliot Morrison in *Admiral of the Open Sea* (Boston: Little,
Brown, 1942), which includes a detailed map of Columbus's daily position based on his own records. Captain Cook's discovery and naming
of Magnetic Island, Queensland, Australia, is described in James G.
Porter's *Discovering Magnetic Island* (1983), which quotes from Cook's
journals as edited by J. C. Beaglehole. This information was provided
to me by Sue Wilkinson, a ranger at Magnetic Island National Park.
James Clark Ross's unsuccessful search for the South Magnetic Pole is
recounted in his *Voyage of Discovery and Research in the Southern and
Antarctic Regions* (London: John Murray, 1847).

In *The Ocean of Truth* (Princeton: Princeton University Press, 1986), an
interesting personal history of research on seafloor spreading and continental drift, H. W. Menard calls paleomagneticians "paleomagicians."
Most geology texts discuss the role of rock magnetism in developing
our present understanding of continental drift. Greater detail can be
found in *Introduction to Geomagnetism* (Amsterdam: Elsevier, 1983), by
W. D. Parkinson, and in *The Earth's Magnetic Field* (New York: Academic
Press, 1983), by R. T. Merrill and M. W. McElhinny. Articles describing
various aspects of the electrodynamic origin of the earth's magnetic field

have appeared in the February 1979, August 1983, May 1988, and December 1989 issues of *Scientific American*.

The book by Merrill and McElhinny mentioned above contains a chapter on lunar magnetism and one on magnetic fields of the sun, planets, and meteorites. Recent articles in *Science* magazine describe detailed findings about magnetic fields at Uranus (4 July 1986), Neptune (5 December 1989), Jupiter (11 September 1992), and the asteroid Gaspra (16 July 1993). An article by Matthew L. Wald in *The New York Times* of June 16, 1991, reviews the effects of the 1989 solar storm. A possible correlation of solar storms and admissions to psychiatric hospitals is reported in Robert O. Becker, *The Body Electric* (New York: William Morrow, 1985), written with Gary Selden.

Magnetic bacteria are described by R. P. Blakemore and R. B. Frankel in *Scientific American* (December 1981) and in the journal *Bioelectromagnetics* (vol. 10, 1989). Frankel and Blakemore are also the editors of *Iron Biominerals* (New York: Plenum, 1989), the proceedings of a conference on the production by living organisms of magnetic iron compounds.

///// 4. Supermagnets

For an overview of the recent achievements and future potential of materials science and engineering, see the 1989 National Research Council report, *Materials Science and Engineering for the 1990s: Maintaining Competitiveness in the Age of Materials*, written by a blue-ribbon committee chaired by P. Chaudhuri and M. Flemings. The song "Material Girl" is from Madonna's album *Like a Virgin* and is copyrighted by Warner Brothers. M. E. Weeks and H. J. Leicester, *Discovery of the Elements*, 7th ed. (Easton, Pa.: Journal of Chemical Education, 1968), R. J. Tayler, *The Origin of the Chemical Elements* (London: Wykeham, 1972), L. H. Aller, *The Abundance of the Elements* (New York: Interscience, 1961), and L. H. Ahrens, *Distribution of the Elements in Our Planet* (New York: McGraw-Hill, 1965), are recommended for readers wanting more background on the chemical elements. The anonymous booklet *On Steel and Iron* (Nuremberg, 1532) is one of many interesting entries in *Sources for the History of the Science of Steel 1532–1786* (Cambridge, Mass.: MIT Press, 1968), edited by Cyril Stanley Smith. Also recommended is "Pathways to Steel" by N. J. van der Merwe and D. H. Avery in volume 70 of *American Scientist* (1982). Because of the importance of the rare-earth metals in modern magnets, D. N. Trifonov's *The Rare-Earth Elements* (New York: MacMillan, 1963), translated from the Russian by P. Basu, is also of interest.

"Permanent Magnets," by Jan F. Herbst (volume 81, 1993, of *American Scientist*), offers an accessible overview of the science and applications of permanent magnets, with emphasis on neodymium-iron-boron magnets. Earlier rare-earth magnets are the subject of E. A. Nesbitt and J. H. Wernick, *Rare Earth Permanent Magnets* (New York: Academic Press, 1973). For permanent-magnet materials prior to rare-earth magnets, and for many of the older applications of permanent magnets, see *Permanent Magnets and Magnetism* (New York: Wiley, 1962), edited by D. Hadfield. A more up-to-date source is R. J. Parker, *Advances in Permanent Magnetism* (New York: Wiley, 1990).

Research in high magnetic fields at MIT's Francis Bitter National Magnet Laboratory is described by L. G. Rubin and P. A. Wolff in the August 1984 issue of *Physics Today*. Also recommended are articles in *Scientific American* on "Intense Magnetic Fields" by H. H. Kolm and A. J. Freeman (April 1965), on "Ultrastrong Magnetic Fields" by Francis Bitter (July 1965), and on "Building World-Record Magnets" by G. Boebinger, A. Passner, and J. Bejk (June 1995). Francis Bitter's autobiography, *Magnets: The Education of a Physicist* (Garden City, N.Y.: Doubleday, 1959), discusses both his development of high-field electromagnets and some of his magnetics research in World War II. See also *Francis Bitter: Selected Papers and Commentaries* (Cambridge, Mass.: MIT Press, 1969), edited by T. Erber and C. M. Fowler. The dedication of Florida's National High Magnetic Field Laboratory is reported in *Physics Today* of December 1994.

///// 5. Superconducting Magnets

The contributions of Ampère, Volta, Ohm, Gauss, Oersted, and other pioneers in the study of electromagnetism are described in S. P. Bordeau, *From Volts to Hertz* (Minneapolis: Burgess, 1982). G. Vidali, *Superconductivity: The Next Revolution?* (Cambridge: Cambridge University Press, 1993), a popularized introduction to the science and applications of superconductivity, includes a discussion of electrical conduction, resistance, and Ohm's law. Other books stimulated by the advent of "high-temperature" superconductivity include C. W. Billings, *Superconductiv-*

ity: From Discovery to Breakthrough (New York: Cobblehill Books, 1991), R. M. Hazen, *The Breakthrough: The Race for the Superconductor* (New York: Ballantine, 1989), and V. D. Hunt, *Superconductivity Sourcebook* (New York: Wiley & Sons, 1989). D. Larbalestier, G. Fisk, B. Montgomery, and D. Hawksworth collaborated on an excellent overview of the applications of high-field superconducting magnets, which appeared in the March 1986 issue of *Physics Today*. (This special issue, devoted to superconductivity on the occasion of the seventy-fifth anniversary of its discovery, was published one month before Bednorz and Müller submitted their "breakthrough" paper to *Zeitschrift für Physik*.) Progress in the application of high-temperature superconductors was reported in *Scientific American* (December 1993).

The political and financial history of the SSC was reviewed in D. J. Kevles, *The Physicists: The History of a Scientific Community in Modern America* (Cambridge, Mass.: Harvard University Press, 1995) and in the October 29, 1993, issue of *Science*. Articles in *Science* also covered the possible discovery of the top quark at the Tevatron (29 April 1994), record-setting fusion energy output at the Princeton tokamak (2 December 1994), and funding problems facing future fusion research (29 April 1994 and 13 January 1995).

///// 6. Inside Magnets and Superconductors

David Jiles, *Introduction to Magnetism and Magnetic Materials* (London: Chapman and Hall, 1991), is more advanced but also more up-to-date than E. W. Lee's *Magnetism*, cited in Chapter 1. B. D. Cullity, *Introduction to Magnetic Materials* (Reading, Mass.: Addison-Wesley, 1972), is a deservedly popular college-level text. Magnetic domains observed by various techniques can be seen in D. J. Craik and R. S. Tebble, *Ferromagnetism and Ferromagnetic Domains* (Amsterdam: North-Holland, 1965), and R. Carey and E. D. Isaac, *Magnetic Domains and Techniques for Their Observation* (New York: Academic Press, 1966). My own "Microstructure and Coercivity in Permanent-Magnet Materials," in *Progress in Materials Science* (Chalmers Volume, 1981), and "The Effects of Metallurgical Variables on Superconducting Properties," *Progress in Materials Science* (volume 12, 1964), focus specifically on structure-sensitive magnetic and superconducting properties. A. McCurrie, *Ferromagnetic Materials: Structure and Properties* (London: Academic Press, 1994), covers a wide range of hard and soft magnetic materials.

///// 7. Attractors, Movers, and Shakers

Most books on magnetism cited in previous chapters describe a variety of magnet applications, as do C. Heck, *Magnetic Materials and Their Applications* (New York: Crane, Russak, 1974), J. K. Watson, *Applications of Magnetism* (New York: Wiley, 1980), and F. N. Bradley, *Materials for Magnetic Functions* (New York: Hayden Book, 1971). But for simple and clear explanations of motors, speakers, transformers, generators, computers, and other devices referred to in this and subsequent chapters, it's hard to beat *The Way Things Work*, cited in Chapter 1.

Refrigerator magnets as collectibles were discussed in *Newsweek* (18 September 1989) and *The Boston Globe* (24 February 1994). Among the many manufacturers of decorative magnets, Funtastic Magnets of Heseria, California, specializes in food magnets, and Acme Magnets of Maplewood, New Jersey, in appliance magnets. Magna-Tel, Inc. of Cape Girardeau, Missouri, and Magnet, Inc. of Washington, Missouri, are major suppliers of promotional magnets. For detailed information on The Gripper, contact Ross E. Meyer of Los Alamos National Laboratory, Los Alamos, New Mexico.

///// 8. AC, RF, TV, and EAS

T. P. Hughes, *Networks of Power: Electrification in Western Society, 1880–1930* (Baltimore: Johns Hopkins University Press, 1983), gives a detailed history of the development of electrical power systems in the United States and Europe. The AC/DC "battle of the currents" is described by T. S. Reynolds and T. Bernstein in *Proceedings of the IEEE* (September 1976) and in more detail by T. P. Hughes in *Harvard Business History Review* (vol. 32, 1958). Magnetic EAS sensors are reviewed by R. C. O'Handley in *Journal of Materials Engineering and Performance* (April 1993), and soft magnetic materials, with emphasis on amorphous metals, are reviewed by G. E. Fish in *Proceedings of the IEEE* (June 1990).

///// 9. Thanks for the Memories

Broad overviews of this field are available in M. Camras, *Magnetic Recording Handbook* (New York: Van Nostrand, 1988), and in *Magnetic Recording Handbook: Technology and Applications* (New York: McGraw-Hill, 1990), edited by C. D. Mee and E. D. Daniel. Useful review articles include "Disk-Storage Technology" by R. M. White in *Scientific American* (August 1980), "Magnetic Information Technology" by M. H. Kryder and A. B. Bortz in *Physics Today* (December 1984), and "Data-Storage Technologies for Advanced Computing" by M. H. Kryder in *Scientific American* (October 1987). *The Big Easy,* produced in 1987 by Kings Road Entertainment, starred Dennis Quaid and Ellen Barkin and is available on videotape, stored far away from any alnico magnets.

The original computer "bug" is reported in J. W. Cortada, *Historical Dictionary of Data Processing Technology* (New York: Greenwood, 1987) and in Ira Flatow, *They All Laughed* (New York: HarperCollins, 1992). The interview with Grace Hopper appears in *Voice of America Interviews with Eight American Women of Achievement,* edited by C. Mompullan (Washington, D.C.: U.S. Information Agency, 1984).

S. Augarten, *Bit by Bit: An Illustrated History of Computers* (New York: Ticknor & Fields, 1984), contains the quote on MIT's Whirlwind and photos of magnetic core memories and many early computers. The March 1990 issue of *MRS Bulletin,* a publication of the Materials Research Society, includes several review articles on the present and future of magnetic recording materials, including both disks and heads.

///// 10. Up with Magnets!

Various approaches to magnetic levitation and suspension are reviewed in B. V. Jayawant, *Electromagnetic Levitation and Suspension Techniques* (London: Edward Arnold, 1981). Magnetic bearings for ultracentrifuges appear in "Ultrahigh-Speed Rotation" by J. W. Beams in *Scientific American* (April 1961). G. R. Polgreen, *New Applications of Modern Magnets* (London: Macdonald, 1966), focuses on the then newly developed hard ferrite magnets and their possible use for maglev trains. Other approaches to maglev trains, including some of the personalities and politics involved, are described in the August 1992 issue of *Scientific American.* J. Vranich, *Supertrains: Solutions to America's Transportation Gridlock* (New York: St. Martin's, 1991), persuasively argues the promise of maglev and other high-speed train systems for the United States.

///// 11. Magnets at War

Hitler's magnetic mines and Allied countermeasures are described in G. W. Gray, *Science at War* (New York: Harper, 1943), and in Bitter's autobiography, cited in Chapter 4. That story from the opposite side is presented by Gerhard Hennig in his privately published *Magnete Sind Uberall* (1993), an autobiography of a German magnet expert, and in *Magnets Are Everywhere* (1995), an English translation of this interesting book. J. P. Baxter, 3rd, *Scientists against Time* (Cambridge, Mass.: MIT Press, 1968), covers magnetic mines, MADs, and other scientific aspects of World War II weaponry, as does T. Hughes, *Battle of the Atlantic* (New York: Dial, 1977).

The story of radar in World War II has been told by Sir Robert Watson-Watt, the "father of radar," in *Three Steps to Victory* (London: Odhams Press, 1957). Other treatments include E. G. Bowen, *Radar Days* (Bristol: Adam Hilger, 1987), J. Nissen with A. W. Cockerill, *Winning the Radar War* (London: Robert Hale, 1989), and D. E. Fisher, *A Race on the Edge of Time* (New York: McGraw-Hill, 1988). The definitive book on the atomic bomb project, R. Rhodes, *The Making of the Atomic Bomb* (New York: Simon & Schuster, 1988), includes a photo of a calutron array. The original "Smyth Report," *Atomic Energy for Military Purposes* (Princeton: Princeton University Press, 1945), by H. D. Smyth, still makes interesting reading; it includes a chapter on the electromagnetic separation of uranium isotopes.

///// 12. Magnets at Play

A strong permanent magnet is among the items that accompany J. Cassidy's *Explorabook* (Palo Alto, CA: Klutz, 1991). Several toys employing magnets are discussed by R. C. Turner in "Physics and Toys," an essay that appears in D. Halliday and R. Resnick, *Fundamentals of Physics,* 3d ed. (New York: Wiley, 1988). Magnetic playing cards and a variety of magnetic puzzles and games are made by the appropriately named Kling Magnetics of Hudson, New York. The magnetic sculptures of crdl, Inc., of Northridge, California, and the "perpetual motion" toys of Andrews Manufacturing of Eugene, Oregon, are targeted more for adults than for children. To build your own magnetic "perpetual motion" toys, see J. Walker's description in "The Amateur Scientist" column in *Scientific American* (March 1982). In A. W. J. G. Ord-Hume, *Perpetual Motion: The History of an Obsession* (New York: St. Martin's, 1977), devices based on magnets appear in chapter 5; the earliest described is from 1269. G. Hennig, in his autobiography cited in the previous chapter, describes many modern-day inventors who come to him with magnetic devices that they believe can produce endless energy. (One was convinced that he was being followed by agents of the international oil cartels, who were out to quash his invention.)

Jonathan Swift's magnetically levitated island of Laputa is critiqued by Isaac Asimov in *The Annotated Gulliver's Travels* (New York: Clarkson Potter, 1980). The article on Dick Tracy in the *Encyclopedia of American Comics* (New York: Facts on File, 1990) claims that Chester Gould "quite truly believed that, as the strip never ceased to declare, the nation that controls magnetism will control the universe." The Magnet Mountain that pulls nails out of ships is quoted from "The Third Kalandar's Tale" in R. F. Burton, *The Arabian Nights' Entertainments or The Book of a Thousand Nights and a Night* (New York: Modern Library, 1932). (Other translations call it "The Third Dervish's Tale" or "Story of the Third Calender, Son of a King.") *Star Trek VI: The Undiscovered Country* was produced in 1991 by Paramount Pictures and is available on videotape. Two X-Men comic books that feature Magneto are volume 1, numbers 1 and 2 (October and November 1991). In the second, Magneto is attacked by missiles "constructed of plastic and molded ceramic composites" to reduce their vulnerability to his magnetic powers (fact 3), but he destroys them with steel I-beams that he can control, using his magnetic forces that act at a distance (fact 4).

The "light and heavy chest" of Robert-Houdin is described in M. Christopher, *The Illustrated History of Magic* (New York: Crowell, 1973). P. Doherty and J. Cassidy, *Magnetic Magic* (Palo Alto, CA: Klutz, 1994), include several simple magic tricks done with permanent magnets. Uri Geller and other deceivers of the gullible are debunked in M. Gardner, *Science: Good, Bad, and Bogus* (Buffalo, NY: Prometheus, 1981; paperback, 1989). Another prominent debunker of pseudoscience is James Randi, professional magician and author of *Flim-Flam!* (Buffalo, NY: Prometheus, 1982).

///// 13. Mesmerism and Magnetic Therapy

The quotation by Jung on Paracelsus is taken from the 15th edition (1989) of *The New Encyclopedia Britannica*. This entry on Paracelsus also includes references to his published writings and several biographies. In *Extraordinary Popular Delusions and the Madness of Crowds* (Boston: L. C. Page, 1932; originally published in 1841), Charles Mackay says of

Paracelsus that "his claim to be the first of the magnetisers can scarcely be challenged." Paracelsus, Mesmer, and other personalities in the history of magnets in medicine appear in "From Thales to Lauterbur, or From the Lodestone to MR Imaging: Magnetism and Medicine," M. R. Maurino's article in *Radiology* (vol. 180, 1991). There are several interesting biographies of Mesmer, of which V. Buranelli, *The Wizard From Vienna* (New York: Coward, McCann & Geoghegan, 1975), is the most balanced. The impact of Mesmerism on the development of hypnotherapy is described in A. Gauld, *A History of Hypnotism* (Cambridge: Cambridge University Press, 1992). For a lively, illustrated review of Mesmer's legacy in America and many other colorful aspects of "alternative medicine," see D. Armstrong and E. M. Armstrong, *The Great American Medicine Show* (New York: Prentice Hall, 1991). H. Schiegl, *Healing Magnetism: The Transference of Vital Force* (York Beach, ME: S. Weiser, 1987), is a modern instructional manual of animal magnetism.

James Graham's Celestial Bed appears in M. Eden and R. Carrington, *The Philosophy of the Bed* (London: Spring Books, 1966). In "Magnetic Healing, Quackery, and the Debate about the Health Effects of Electromagnetic Fields," published in *Annals of Internal Medicine* (March 1993), R. M. Macklis discusses C. J. Thacher and other "magnetic quacks" (and argues that the shady history of magnets in medicine has hampered serious study of the effects of electromagnetic fields on health). L. Johnson, *Magnetic Healing and Meditation* (San Francisco: White Elephant Monastery, 1988), and B. Payne, *The Body Magnetic*, 5th ed. (Boulder, CO: self-published, 1991), describe some aspects of modern magnetotherapy.

///// 14. Medicine and MRI

An article by F. E. Luborsky et al. in the November 1964 issue of the *American Journal of Roentgenology, Radium Therapy and Nuclear Medicine* describes the on-off magnetic device pictured removing a safety pin and includes references to earlier work on magnetic removal of foreign bodies. Other articles on magnets in medicine include: on PEMF, C. A. L. Bassett, *Critical Reviews in Biomedical Engineering* (vol. 17, 1989); interstitial hyperthermia, M. Sato et al., *IEEE Transactions on Magnetics* (vol. 29, November 1993); catheters, S. K. Hilal et al., *Radiology* (vol. 113, December 1974); IUDs, B.R. Livesay et al., *Proceedings of the 9th International*

Conference on Rare-Earth Magnets and Their Applications (Bad Honnef, Germany: Deutsche Physikalische Gesellschaft, 1987); embolization, J. W. Barry et al., *Radiology* (vol. 138, February 1981); bioseparation, J. Ugelstad et al., *Progress in Polymer Science* (vol. 17, 1992); cyclotrons and radiation therapy, H. G. Blosser, *Physics Today* (October 1993); and cyclotrons and PET, J. Gordon, *Physics World* (June 1994). Prostheses and other uses of rare-earth permanent magnets in medicine are reviewed by E. A. Smith in *Elements* (February/March 1994).

J. F. Schenck reviews "Magnetic Resonance in Medicine" in volume 9 of the *Encyclopedia of Physical Science and Technology*, 2d ed. (New York: Academic Press, 1992). See also T. F. Budinger and P. C. Lauterbur, *Science* (October 1984), G. E. Pake, *Physics Today* (October 1993), and A. D. Elster (with S. F. Handel and A. M. Goldman), *Magnetic Resonance Imaging: A Reference Guide and Atlas* (Philadelphia: Lippincott, 1986). M. E. Raichle (*Scientific American*, April 1994) and R. Turner and P. Jezzard (*Physics World*, August 1994) show how brain functions are studied by PET and "functional" MRI. (Most often MRI is used to learn about anatomy and pathology. Functional MRI can also yield information about blood flow, blood oxygenation, metabolism, and neuronal activity.)

///// 15. Biomagnetism

Early work on biomagnetic fields was reviewed by David Cohen in *Physics Today* (August 1975). See also P. Gwynne, "The Medical Promise of Personal Magnetism" (*Technology Review*, August/September 1985), and R. Crease, "Biomagnetism Attracts Diverse Crowd" (*Science*, 8 September 1989). M. Hamalainen et al., in the April 1993 issue of *Reviews of Modern Physics*, presented a comprehensive review of magnetoencephalography. For SQUIDs, see J. Clarke's article in the August 1994 *Scientific American*.

The effects of human exposure to 4 tesla steady fields are described in J. F. Schenck et al., *Medical Physics* (July/August 1992), and in volume 649 of the *Annals of the New York Academy of Sciences* (March 31, 1992). The latter includes the quotation cited from 1892. A wide variety of experiments are reported in two conference proceedings: *Biological Effects of Magnetic Fields*, edited by M. F. Barnothy (New York: Plenum, 1964), and *Biophysical Effects of Steady Magnetic Fields*, edited by G. Maret (Berlin: Springer, 1986). Books raising concern about EMF and health include P. Brodeur, *Currents of Death: Power Lines, Computer Terminals and the Attempt to Cover Up Their Threat to Your Health* (New York: Simon & Schuster, 1989), E. Sugarman, *Warning: The Electricity Around You May Be Hazardous to Your Health* (New York: Simon & Schuster, 1992), and R. O. Becker, *Cross Currents: The Promise of Electromedicine, The Perils of Electropollution* (Los Angeles: Jeremy Tarcher, 1990). On the other side of the controversy are E. L. Carstensen, *Biological Effects of Transmission Line Fields* (New York: Elsevier, 1987), W. R. Bennett, Jr., *Health and Low-Frequency Electromagnetic Fields* (New Haven: Yale University Press, 1994), and an article by Bennett in *Physics Today* (April 1994) and one by Gary Taubes in *The Atlantic Monthly* (November 1994).

Cole Porter's "Let's Do It," performed most notably by Noel Coward, is copyrighted by Warner Brothers. To read more about magnetic sensitivity in birds and bees, a good place to start is J. L. Gould's article in *American Scientist* (May–June 1980). B. R. Moore, in *Proceedings of the National Academy of Science* (July 1988), discusses Keeton's experiments on pigeons, including later results indicating that "no overall effect has been shown." A multi-authored review of bird navigation appeared in *Experientia* (vol. 46, 1990). For honeybees, J. L. Kirschvink et al., in *American Zoology* (vol. 31, 1991), describe magnetic conditioning experiments; M. W. Walker and M. E. Bitterman, *Journal of Experimental Biology* (vol. 141, 1989), attached magnets to bees' abdomens; and C.-Y. Hsu and C.-W. Li, *Science* (1 July 1994), found magnetite particles in cells connected to bees' nervous systems. Experiments on hatchling sea turtles were done by K. J. Lohmann and C. M. F. Lohmann, *Journal of Experimental Biology* (vol. 190, 1994), and reviewed by L. Seachrist in *Science* (29 April 1994). Baker's experiments on university students appeared in *Science* (31 October 1980) and are summarized in R. R. Baker, *Human Navigation and Magnetoreception* (Manchester: Manchester University Press, 1989). J. L. Kirschvink found magnetite crystals in the human brain (*Science*, 15 May 1992) and suggested, in J. L. Kirschvink et al., *Bioelectromagnetics* (supplement 1, 1992), that magnetite in human tissues may provide a mechanism for biological effects of weak magnetic fields.

///// 16. Source of the Force

The Clark Kent/Superman analogy comes from K. Amdahl, *There Are No Electrons: Electronics for Earthlings* (Arvada, CO: Clearwater, 1991), a very lively and very offbeat tutorial on electronics. Also lively and offbeat is L. Gonick and A. Huffman, *The Cartoon Guide to Physics* (New York: HarperCollins, 1990). Einstein's teenage paper on magnetic fields can be found in *The Collected Papers of Albert Einstein, Volume 1: The Early Years: 1879–1902* (Princeton: Princeton University Press, 1987). Edward Purcell's remarks on magnetism as an "engineering application of relativity," and his discussion of the Leigh Page article, appear in *Some Strangeness in the Proportions: A Centennial Symposium to Celebrate the Achievements of Albert Einstein,* edited by H. Woolf (Reading, MA: Addison-Wesley, 1980). Feynman's BBC interview and Gleick's quote about quantum mechanics and materials scientists are from J. Gleick, *Genius: The Life and Science of Richard Feynman* (New York: Random House, 1992), and the quote about the discovery of the wave equation for electrons is from W. Moore, *Schrödinger: Life and Thought* (Cambridge: Cambridge University Press, 1989).

R. P. Feynman, *QED: The Strange Theory of Light and Matter* (Princeton: Princeton University Press, 1985), is a highly readable presentation of a very abstruse but successful theory. The development of QED is described in S. S. Schweber, *QED and the Men Who Made It: Dyson, Feynman, Schwinger, and Tomonaga* (Princeton: Princeton University Press, 1994). A very entertaining treatment of QED, QCD, and advances in particle physics and cosmological physics is available in L. Lederman (with D. Teresi), *The God Particle: If the Universe Is the Answer, What Is the Question?* (New York: Delta, 1993).

////// ACKNOWLEDGMENTS

I am very grateful to various friends and colleagues who read all or part of the manuscript and made helpful suggestions, especially David Paul and Fred Rothwarf, who had the courage and patience to tackle every chapter. Others who kindly reviewed one or more chapters include Ami Berkowitz, Larry Bennett, David Cohen, Richard Frankel, Howard Hart, Mark Kryder, Alan Lightman, Ted Madden, Joseph Vranich, and George Wise. Fred Jones, Luo Yang, Motofumi Homma, and the late Rollin Parker provided very helpful advice by mail, and Gerhard Hennig was kind enough to give me copies of the German and English versions of his autobiographical book, *Magnets Are Everywhere*. My daughter Joan, an expert editor, greatly improved the writing in my first draft, and my wife Sherry, a former history major and now a busy university president, provided helpful comments from the perspective of a nontechnical reader. Despite help from all the aforementioned, some errors undoubtedly remain, and readers are encouraged to call them to my attention.

Thanks are also due my agent Ed Knappman, the first from the world of publishing who showed enthusiasm for my project, and my editor Michael Fisher, who soon thereafter became the second. Michael, my anonymous reviewers, and later Kate Schmit, my copy editor, suggested many changes that made the book more readable and understandable.

I owe special thanks to Charlie Bean, the late Joe Becker, and other former colleagues at General Electric who introduced me to the fascinating world of magnets. Finally, my thanks to the MIT students in my freshman seminar, "The Magic of Magnets and Superconductors," and to those in my other courses in the Department of Materials Science and Engineering, who helped me judge the relative attraction and the relative difficulty of various aspects of the science and technology of magnets.

Graphs and illustrations were prepared by Dmitri Schidlovsky Illustration, Sea Cliff, New York. Previously published figures and other photographs are reproduced here with thanks:

Figure 2.1 From Joseph Needham, *Science and Civilization in China,* vol. 4, pt. 2 (Cambridge, 1962); reprinted with the permission of Cambridge University Press.

Figure 2.2 Photo courtesy of National Maritime Museum, Greenwich, London.

Figure 3.3 From L. D. Leet, M. E. Kauffman, and S. Judson, *Physical Geology,* 7th ed. (Englewood Cliffs, N.J.: Prentice-Hall, 1987).

Figure 3.4 From M. W. McElhinny, *Palaeomagnetism and Plate Tectonics* (Cambridge, 1973); reprinted with the permission of Cambridge University Press.

Figure 3.5 Adapted from R. T. Merrill and M. W. McElhinny, *The Earth's Magnetic Field* (New York: Academic Press, 1983).

Figure 3.6 From D. Halliday and R. Resnick, *Fundamentals of Physics,* 3d ed. (New York: Wiley, 1988); reprinted by permission of John Wiley & Sons, Inc. Photo courtesy of R. Blakemore and N. Blakemore.

Figure 4.2 Photo courtesy of Larry Rubin, Francis M. Bitter National Magnet Laboratory, Massachusetts Institute of Technology, Cambridge, Massachusetts.

Figure 5.3 Photo courtesy of Fermilab Visual Media Services.

Figure 6.6 Photo courtesy of G. J. Dolan, University of Pennsylvania.

Figure 7.1 Photo by Fred H. Rick, Los Alamos National Laboratory.

Figure 7.2 Diagram courtesy of Magnequench (a business unit of Delco Remy, Division of General Motors), Anderson, Indiana.

Figures 8.3 and 8.4 Diagrams and photo courtesy of Applied Signal Amorphous Metals, Parsippany, New Jersey.

Figures 9.1 and 9.3 Adapted with permission from M. Camras, *Magnetic Recording Handbook* (New York: Van Nostrand Reinhold, 1988).

Figure 10.2 Photo by Donna Coveney; reprinted with permission from the M.I.T. News Office.

Figure 10.3 Adapted from R. J. Parker, *Advances in Permanent Magnetism* (New York: Wiley, 1990); reprinted by permission of John Wiley & Sons, Inc.

Figure 10.4 Photo courtesy of General Electric Meter & Control, Somersworth, New Hampshire.

Figure 10.5 From B. V. Jayawant, *Electromagnetic Levitation and Suspension* (London: Edward Arnold, 1981).

Figure 10.6, top Photo courtesy of the Railway Technical Research Institute, Shinjuku Office, Tokyo.

Figure 10.6, bottom Photo courtesy of Transrapid International, Munich.

Figure 11.1, top From E. G. Bowen, *Radar Days* (Bristol: IOP Publishing, 1987). Reprinted with permission from IOP Publishing.

Figure 11.1, bottom From J. Nissen and A. W. Cockerill, *Winning the Radar War* (Toronto: Macmillan, 1987); copyright 1987, published by Macmillan Canada. Reprinted by permission of Jack Nissen.

Figure 12.1 "Sculpture" by crdl, Inc., North Hollywood, California.

Figure 12.2 From R. C. Turner, "Physics and Toys," in D. Halliday and R. Resnick, *Fundamentals of Physics,* 3d ed. (New York: Wiley, 1988); reprinted by permission of John Wiley & Sons, Inc.

Figure 12.3 ROBOTMAN reprinted by permission of NEA, Inc.

Figure 13.1 Reproduced from E. G. Gartrell, compiler, *Electricity, Magnetism, and Animal Magnetism: A Checklist of Printed Sources, 1600–1850* (Philadelphia: American Philosophical Society, 1975), with permission from the American Philosophical Society.

Figure 13.2 Drawing by Crawford; copyright 1993 The New Yorker Magazine, Inc.

Figure 14.1 Photo No. RL-40, 313, Research Information Section, General Electric Research Laboratory, Schenectady, New York.

Figures 14.2, left, and 14.5 Photo courtesy of General Electric Medical Systems, Milwaukee, Wisconsin.

Figure 14.2, right Photo courtesy of Harvey Cline and William Lorensen, General Electric Company, Schenectady, New York.

Figure 14.4 Photo courtesy of Sumitomo Special Metals Co., Osaka.

Figure 15.1 Photo courtesy of David Cohen, Massachusetts Institute of Technology, Cambridge, Massachusetts.

Figure 15.2 Photo courtesy of CTF Systems Inc., Port Coquitlam, British Columbia, Canada.

Figure 16.1 From Larry Gonick and Art Huffman, *The Cartoon Guide to Physics* (New York: HarperCollins, 1990); copyright 1990 by Lawrence Gonick and Arthur Huffman. Reprinted by permission of HarperCollins Publishers, Inc.

Figure 16.2 Photo courtesy of the Stanford Linear Accelerator Center, Stanford, California.

Dave Barry excerpt on page 265 From "Science IQ," syndicated June 1994; used with permission.

///// INDEX

hydrogen, 49, 86, 227–228
hyperthermia, 221
hypnosis, 208
hysteresis loop, 97

IBM, 82, 145, 150
Ibn Hazm, 15
implosion magnets, 60–61, 63
inclination, 29, 38
indium, 69
induction, electromagnetic (fact 8), 9,
    12, 257–260; applications, 124–125,
    128, 143, 157–158, 240; currents in the
    body, 220, 248
inductor, 131
infrared, 129–130
insulators, 53, 67
integrated circuit, 147, 148
International Thermonuclear
    Experimental Reactor (ITER), 82
intrauterine device (IUD), 222
Iowa State University, 54
iron, 13–14, 18–21, 48–49, 87–88,
    280–281; and steel, 50–51, 280–281; in
    permanent magnets, 51, 53, 55; as
    soft magnetic material, 100, 134–135;
    in human body, 211, 239
Iron Curtain, 198
iron ore, 23–24
iron oxide, 17–18, 49, 53, 136; in
    magnetic recording, 141, 142, 149, 151
iron-silicon, 135, 138
isotopes, 184, 222–223
isotope separation, 164, 183–185

Japanese National Railways, 168–170
Jaws, 6, 58, 112, 161
Jefferson, Thomas, 208, 210
Johnson, Larry, 214
Jonson, Ben, 15, 28
Jung, Carl, 203

Kamerlingh Onnes, Heike, 68, 69–70,
    79, 103
Keeton, William, 251–252, 253
Kemmler, William, 124, 127
Kent, Clark, 257, 258

King, Rodney, 144
Kirschvink, Joseph, 249
Klingons, 195
Knight, Gowen, 211
Kolm, Henry, 171

Lafayette, Marquis de, 204, 206–207
Lauterbur, Paul, 232
Lavoisier, Antoine, 207
Lawrence, Ernest, 184
lead, 48, 69, 103
leukemia, 249
life, magnetism of, 240
light, 129–130, 256
lightning, 21–22
Limbaugh, Rush, 130
linear motors, 116–117
Little Boy, 183
little science, 75
lodestones, 13–23, 28, 29–30, 50, 51, 53,
    85, 96, 255
Logachev, A., 24
logarithmic scale, 63–64, 101, 130
Los Alamos National Laboratory, 60,
    110
Louis XVI, 207
"loving stones," 14–15, 212
Lucretius, 101

madcats, 176
Madonna, 47, 278
maghemite, 18, 96
magic, 190, 196–197
maglevity, 194
maglev trains, 116, 165–172
Magnalawn, 199
Magnarail, 167
Magneplane, 171
Magnequench, 57
magnet, origin of word, 14
"magnet and churn," 25–26, 85
Magnet Mantra, 214
Magnet Mountain, 23, 194
magnetic anisotropy, 98–102, 131
magnetic anomalies, 23
magnetic anomaly detectors (MADs),
    23–24, 175, 176

niobium-tin, 70, 71, 72, 73, 81, 82, 84, 104, 165

niobium-titanium, 71, 72, 79, 80, 81, 82, 104, 233

nitrogen, liquid, 83, 105

Nobel Prize, 51, 83, 226, 262

Norman, Robert, 29

North Carolina, University of, 254

northern lights, 42, 76, 77

nuclear magnetic resonance, 227, 229–231

nucleus, 48, 63, 76, 227

Oak Ridge, 54, 82, 185

Oersted, Hans Christian, 52, 65, 86, 256

ohm (unit), 65

Ohm, Georg Simon, 65

Onnes, Heike Kamerlingh, 68, 69–70, 79, 103

optical recording, 152

orbital magnetism, 86

Oriental Medical Supplies, 214–215

oxides, 17–18, 53, 82–83, 104, 107

Page, Leigh, 258

paleomagnetism, 34

Palmer, Daniel David, 208

Paracelsus, 202–203

paramagnetism, 87

paranormal, 201

particle physics, 76, 268–269

Patience (Gilbert & Sullivan), 24–26

Payne, Buryl, 216–217

periodic table, 47, 48

permalloy, 100, 135

permanent (hard) magnets, 3–5, 10, 11, 18–20, 63, 90, 134; materials, 50–57; origin of properties, 95–101

permeability, 92, 98, 102, 135

perpetual motion, 191–193, 293–294

Philips Research Laboratories, 166

photon, 262, 263

photon, virtual, 265–266, 267, 268

pigeons, 251–252

pixel, 224–226

placebo effect, 217

planets, 40–41

plasma, 81

plastics (polymers), 4, 57, 67, 107, 277

plate tectonics, 24, 35

Plato, 13

Pliny, 14, 193

plutonium, 185

poles, magnetic, 10, 89, 92, 109, 115, 156, 217, 264; labeling, 10, 33, 217; earth's, 29–35, 38

Polgreen, Geoffrey, 167

poltergeists, 74, 201

Porta, Giambattista della, 22

Porter, Cole, 250

positron, 80, 223, 263

positron emission tomography (PET), 222–223, 224, 226, 227, 232, 245, 246, 263

Poulsen, Valdemar, 139–141

Powell, Gordon, 166

power, 124

power lines, 124, 129–130; and health, 247, 248–249

precession, 229–231

prefixes (for numerals), 62

processing, 46, 71, 95, 136–138, 152, 278, 279, 280

promotional magnets, 108

prospecting, magnetic, 23–24

prostheses, 221

protons, 48, 66, 78–79, 80, 227–229

psychics, 190, 197–198

psychokinesis, 197–198, 200, 201

psychosomatic medicine, 208

Ptolemy II, 193

pulsed electromagnetic fields (PEMFs), 220–221

pulsed magnets, 60–61, 63, 272

Purcell, Edward, 260

quadrupole magnets, 78–79

Quaid, Dennis, 143

quantum chromodynamics (QCD), 268, 269

quantum electrodynamics (QED), 265–269, 282